Contents

CHAPTER ONE

Introduction

I t is becoming more and more obvious that we live in a technologically advanced world as we progress through the twenty-first century. From the moment we wake up until the moment we go to bed, we interact with countless innovations that have shaped every aspect of our lives. The pace at which progress is being made is nothing short of remarkable, and it is crucial for us to understand and appreciate the impact these advancements have on our society.

This book aims to delve into the vast realm of science and technology, exploring its evolution, challenges, and potential. We will witness how these fields have transformed over time, moulding the way we live, work, and communicate. From the humble beginnings of ancient discoveries to the cutting-edge breakthroughs of today, we will trace the roots of innovation and see how it has shaped human civilisation.

Humanity's thirst for knowledge has fuelled the march of progress since ancient times. In the realms of science and technology, we have seen remarkable achievements throughout history. From the invention of the wheel to the discovery of fire, our ancestors planted the seeds of innovation that continue to flourish today.

The Mesopotamians, known for their advancements in mathematics and astronomy, developed the earliest forms of writing and record-keeping. The Egyptians, with their monumental pyramids and impressive irrigation systems, showcased their engineering prowess. The Greeks, masters of reason and philosophy, laid the foundation for the scientific enquiry we know today. These civilisations not only made ground breaking discoveries but also provided a roadmap for future generations to build upon.

With the explosion of the Industrial Revolution in the 18th century, science and technology entered a new era of exponential growth. Steam power, mechanisation, and the assembly line revolutionised manufacturing, leading to the birth of the modern world. The introduction of electricity brought about a paradigm shift, powering homes, cities, and industries, while advancements in transportation made the world more connected than ever before.

In the late 19th and early 20th centuries, the fields of physics and chemistry underwent extraordinary transformations. The discovery of radioactivity by scientists like Marie Curie and the development of the theory of relativity by Albert Einstein revolutionised our understanding of the universe. These breakthroughs paved the way for innovations such as X-rays, nuclear power, and the development of the atomic bomb, forever changing the course of human history.

As the 20th century unfolded, humanity witnessed unparalleled leaps forward in science and technology. The discovery of the double helix structure of DNA by Watson and Crick revolutionised our understanding of genetics, laying the groundwork for advances in medicine. The computer, once a primitive machine, evolved into a powerful tool capable of processing vast amounts of information, giving rise to the digital age. The advent of the internet further connected the world, ushering in an era of unparalleled communication and information sharing.

One of the most powerful forces that science

and technology have unleashed upon us is artificial intelligence. AI, a field that seeks to create intelligent machines capable of mimicking human cognitive processes, has seen tremendous growth in recent years. Machine learning algorithms have enabled computers to analyse vast amounts of data and make predictions with unprecedented accuracy. From voice assistants in our smartphones to self-driving cars, AI has become an integral part of our everyday lives, transforming industries and reshaping the job market.

In the realm of communication, the advances made in the past few decades are truly exceptional. The internet, which started as a small network connecting a few research institutions, has grown into a global phenomenon that has revolutionised the way we communicate, work, and access information. Social media platforms have connected people from all corners of the globe, fostering new forms of virtual communities and online collaboration. Virtual reality and augmented reality have blurred the line between the physical and digital worlds, immersing users in interactive experiences that were once only imaginable in science fiction.

The medical field has also experienced a profound transformation, with advancements in technology revolutionising healthcare. From DNA sequencing and gene editing to robotic surgery and telemedicine, the possibilities seem endless. Personalised medicine, using a patient's genetic information to tailor treatments, is becoming a reality, allowing for more targeted and effective therapies. The age of precision medicine is dawning, promising to improve outcomes and reshape the healthcare landscape.

Moreover, we cannot overlook the pressing issue of energy and sustainability. As the earth's resources dwindle, we must find innovative solutions to ensure a sustainable future.

Renewable energy sources like solar, wind, and tidal power hold the key to reducing our dependence on fossil fuels and mitigating the impact of climate change. The development of smart grids, energy storage solutions, and efficient buildings is transforming the way we produce and consume energy, creating a more sustainable and resilient world.

The boundaries of exploration have also expanded in recent years, and space has become an area of immense innovation. From the monumental achievement of landing humans on the moon during the Apollo missions to the ongoing exploration of Mars and beyond, space agencies and private companies are pushing the limits of human exploration. The potential for extraterrestrial life and the colonisation of other planets have captured our imaginations, inviting us to dream of a future where humans become a multiplanetary species and unlock the secrets of the universe.

Transportation is another field that has seen tremendous progress, with the dream of flying cars and high-speed travel inching closer to reality. Electric vehicles are becoming increasingly prevalent, with advancements in battery technology offering range and efficiency improvements. The development of autonomous vehicles has the potential to revolutionise transportation, making it safer, more efficient, and transforming urban landscapes. The hyperloop concept, with its promise of ultra-fast, low-friction travel, is capturing the attention of engineers and innovators worldwide.

However, as we embrace these advancements, we must also confront the ethical dilemmas they present. The power of science and technology is immense, and with great power comes great responsibility. Issues such as privacy, data security, and the ethical use of AI must be navigated carefully. The potential for job displacement and growing inequality also demands our attention, as the benefits of technological progress must be shared equitably.

Lastly, education is an arena that has undergone a massive shift, as the digital revolution has transformed the way we learn. Online platforms and educational apps are a supplement to, and occasionally a replacement for, traditional classrooms. Distance learning and virtual classrooms have become the norm for many, offering access to education anytime, anywhere. Lifelong learning is now more accessible than ever, enabling individuals to upskill and adapt to the changing demands of the workforce.

In this book, we will embark on an exciting journey through the realms of science and technology, diving deep into their evolution, challenges, and potential. We will witness the remarkable progress that has been made and ponder the potential challenges and opportunities that lie ahead. As we navigate this complex landscape, it is essential to stay informed, critical, and open-minded. Only by understanding the world we live in can we fully appreciate the wonders and possibilities that science and technology offer us.

CHAPTER TWO

The Evolution of Science and Technology

S cience and technology have profoundly shaped the world we live in today. From the earliest discoveries of fire and tools to the modern wonders of artificial intelligence and space exploration, the evolution of science and technology has been a fascinating and complex journey that has revolutionised every aspect of human life.

The journey began thousands of years ago when our ancestors first began to observe and understand the natural world around them. Through curiosity and ingenuity, they made remarkable discoveries that laid the foundation for future advancements. The invention of the wheel, for example, revolutionised transportation and trade, enabling the movement of goods and people over long distances. Alongside practical inventions, early civilisations also sought to understand the cosmos. The Egyptians studied the stars and developed a complex calendar based on their observations, while the Mayans in Central America constructed intricate calendars that accurately predicted astronomical events.

In ancient times, different civilisations made significant contributions to various scientific fields. The Greeks, for instance, with their philosophy and scientific enquiry, laid the groundwork for Western science. Not only did philosophers such as Socrates, Plato, and

Aristotle ponder abstract notions of the universe, but they also made ground breaking discoveries in biology, physics, and astronomy. It was the Greeks who proposed the concept of atoms, suggested that the Earth was a sphere, and accurately calculated the circumference of the Earth. Similarly, the Chinese, with their long history of technological innovations, made remarkable advancements in various fields. They invented items like paper, printing, and the compass, which significantly influenced the world.

The scientific revolution of the 17th century marked a turning point in the history of science. During this time, thinkers like Galileo Galilei, Johannes Kepler, and Isaac Newton proposed revolutionary theories and laws that challenged existing beliefs and shaped modern science. Galileo's use of a telescope allowed him to provide evidence for the heliocentric model of the solar system, overturning the widely-held geocentric view. Kepler

mathematically described the motions of celestial bodies, laying the foundation for Newton's laws of motion and universal gravitation.
Newton's breakthrough theories not only ex-
plained the motion of objects on Earth but
also provided a mathematical framework for
understanding the motion of planets and stars.
Moreover, Francis Bacon's concept of the scientific method, a systematic approach to inves-
tigating the natural world through observation
and experimentation, became the cornerstone
of scientific enquiry.

The scientific revolution was closely inter-
twined with the Industrial Revolution, a period of rapid advancements in technology and
industry that began in the 18th century. Inventions such as the steam engine, telegraph,
and electricity transformed industries and propelled societies into a new era. The ability to
harness and control electricity led to the invention of electric lighting, revolutionising the
way we live and work. The harnessing of steam

power led to the development of trains and ships, revolutionising global trade and transportation. These advancements not only improved the quality of life but also brought about profound social and economic changes.

The 19th and 20th centuries witnessed further significant scientific discoveries and technological breakthroughs that shaped the modern world. The discoveries of DNA and the theory of evolution revolutionised our understanding of life and its origins. The development of vaccines and antibiotics helped control and eradicate deadly diseases, leading to significant improvements in public health. The invention of the telephone and later the internet revolutionised communication, bringing people closer together and enabling instant global connectivity. The development of computers and the internet ushered in the digital age, enabling rapid processing of information and fostering innovation in various fields.

Today, we find ourselves in the midst of a technological revolution that is reshaping every aspect of our lives. Artificial intelligence (AI) has reached new heights, allowing machines to learn, reason, and make decisions with unprecedented capabilities. Robotics and automation have revolutionised industries, leading to greater efficiency and productivity. The internet and digital communication have connected the world like never before, enabling instant access to information and fostering global collaboration across borders and continents.

As the pace of scientific and technological advancements accelerates, the possibilities for human progress seem limitless. Advancements in medicine and healthcare have extended our lifespan and improved our quality of life, tackling diseases and finding treatments that were once unimaginable. The mapping of the human genome has unlocked the potential for personalised medicine, tailoring treatments to the individual. Moreover, breakthroughs in renewable energy, nanotechnology, and biotechnology hold the promise of addressing global challenges such as climate change, food scarcity, and disease outbreaks.

However, amidst this rapid progress, we face numerous challenges and ethical dilemmas. The responsible use of technology is critical to ensure its benefits are distributed equitably and do not exacerbate existing inequalities. Ethical considerations regarding privacy, security, and the impact of automation on employment require careful thought and regulation. The preservation of our environment is crucial as we navigate the consequences of technological advancements and strive for a sustainable future.

In conclusion, the evolution of science and technology has been an extraordinary human endeavour that has transformed every aspect of our world. From the early discoveries of our ancestors to the cutting-edge advancements of today, science and technology have driven progress, sparked innovation, and enriched our lives. As we forge ahead, we must navigate the ethical, social, and environmental challenges that arise and ensure that science and technology are harnessed for the greater good of humanity. Through responsible and thoughtful advancement, we can create a future where science and technology coexist harmoniously with a thriving society and a sustainable environment, benefiting all humankind.

CHAPTER THREE

The Power of Artificial Intelligence

Artificial Intelligence (AI) has emerged as one of the most transformative technologies of our time. The concept of AI involves the development of intelligent machines that can think, learn, and perform tasks typically requiring human intelligence. With advancements in computing power, algorithmic breakthroughs, and the availability of vast amounts of data, AI has become increasingly sophisticated, revolutionising various industries and aspects of our daily lives.

One of the key strengths of AI lies in its ability to process and analyse massive amounts of data at unprecedented speed. This capability has opened up new possibilities in fields such as healthcare, finance, transportation, and entertainment. In the healthcare industry, AI algorithms can analyse medical records, genomic data, and imaging data to identify patterns and predict outcomes. These insights help doctors make more accurate diagnoses, tailor treatment plans to individual patients, and even discover new therapies. Moreover, AI-powered robots and devices can perform complex surgeries with enhanced precision and assist in remote healthcare service delivery in underserved areas.

In the finance sector, AI has revolutionised the way we handle investment decisions and financial management. With the ability to process large volumes of market data in real-time, AI algorithms can identify trends and patterns, helping investors make informed decisions and optimise their portfolios. AI-powered virtual assistants have also become prevalent in the finance industry, providing personalised financial advice and assisting customers with their financial planning and budgeting.

The impact of AI is not limited to these sectors alone. In the realm of transportation, AI is paving the way for autonomous vehicles. Self-driving cars equipped with AI algorithms can analyse their surroundings, anticipate road conditions, and make split-second decisions to ensure safety and efficiency. This technology has the potential to reduce accidents and traffic congestion, as well as provide mobility solutions for populations unable to drive, such as the elderly or disabled. The integration of AI in transportation systems also opens up new opportunities for optimising logistics, reducing emissions, and enhancing the overall efficiency of complex supply chains.

Moreover, AI has revolutionised the way we interact with technology and the Internet. Natural language processing capabilities enable voice assistants like Siri and Alexa to understand and respond to human commands, making it easier to access information and perform tasks using just our voices. AI algorithms power recommendation systems that personalise our online experiences, from suggesting movies and music to predicting our shopping preferences. Furthermore, AI has facilitated advancements in cybersecurity, detecting and preventing cyber threats in real time, safeguarding our digital lives.

However, the rapid advancement of AI also raises important ethical concerns. As AI becomes more autonomous and capable of making decisions, the question of accountability and responsibility arises. There are concerns about biases in AI algorithms, potential job displacement due to automation, and the impact of AI on privacy and security. Various stakeholders, including governments, researchers, and technology companies, are now grappling with these challenges to ensure the responsible development and use of AI.

The potential of AI to address some of the most pressing challenges of our time is immense. In the face of climate change, AI can help optimise energy consumption, enhance predictions of severe weather events, and enable more efficient resource allocation. In healthcare, AI has the potential to revolutionise drug discovery, personalise treatment approaches, and accelerate medical research. Additionally, AI can play a crucial role in enhancing education by providing personalised learning experiences, adapting to individual students' needs, and providing students with real-time feedback.

AI also holds tremendous potential for social impact, such as aiding disaster response efforts by quickly analysing data to identify affected areas and optimise supply chains. It can also contribute to the preservation of cultural heritage by using AI algorithms to restore and digitise fragile artefacts or reconstruct lost historical sites.

As AI continues to evolve, it is essential that we embrace its possibilities while navigating the ethical complexities it presents. We must prioritise transparency and accountability in AI systems, ensuring that they are developed and used in a fair and unbiased manner. This involves addressing issues like data privacy and algorithmic transparency and ensuring AI benefits all members of society, irrespective of their gender, race, or socio-economic background.

The power of artificial intelligence holds immense potential, and it is up to us as individuals, policymakers, and global citizens to harness it for the betterment of society. By fostering collaboration among multiple stake-holders, we can ensure equitable access to AI technologies, mitigate the risks associated with AI deployment, and foster ethical practises. By doing so, we can capitalise on the transfor-mative power of AI, not only for economic growth but also for addressing societal chal-lenges and creating a future that benefits all of humanity.

CHAPTER FOUR

Communication Breakthroughs: From the Internet to Virtual Reality

I n this chapter, we will delve even deeper into the remarkable advancements in communication technology that have transformed the way we connect, share information, and experience the world around us. From the invention of the internet to the emergence of virtual reality, these breakthroughs have revolutionised the way we communicate and interact with each other, shaping our society in profound ways.

The internet, often referred to as the "information superhighway," has become an indispensable part of our daily lives. Its origins can be traced back to the 1960s when researchers sought to create a decentralised network that could survive a nuclear attack. This led to the development of packet-switching, a method of dividing and transmitting data across networks, which formed the foundation of the internet we know today.

Initially, the internet was primarily used by academics and scientists to exchange research findings and collaborate across institutions. However, with the introduction of the World Wide Web in 1989 by Tim Berners-Lee, the internet became accessible to the general public. This marked a turning point, leading to its exponential growth and the birth of a digital revolution.

The internet has democratised access to information, breaking down barriers to knowledge and education. With a few clicks, we can explore vast amounts of information on almost any topic, from historical records to scientific research. This easy access to information has empowered individuals, enabling lifelong learning and fostering a culture of curiosity and knowledge-sharing.

Email revolutionised the way we communicate, replacing traditional letter-writing with near-instantaneous electronic messages. We no longer had to wait days or weeks for a response, as emails could be sent and received within seconds across the globe. This transformed the way businesses operate, facilitating efficient communication between teams, clients, and partners.

Instant messaging and chat platforms such as AOL Instant Messenger, MSN Messenger, and later, WhatsApp and Facebook Messenger, further accelerated real-time communication. We could now chat with friends and loved ones in different time zones, share photos and videos, and even have group conversations with multiple participants.

The rise of social media platforms like Facebook, Twitter, and Instagram brought about a new era of connecting and sharing. People could now broadcast their thoughts, opinions, and experiences to a wide audience, fostering new forms of social interaction. These platforms gave rise to the concept of "virtual communities," where individuals with similar interests and aspirations could connect and form online relationships.

However, the rapid growth of social media also brought forth concerns regarding privacy, cyberbullying, and the spread of misinformation. Striking a balance between the benefits of connectivity and the need for responsible online behaviour became an ongoing challenge. Efforts to combat these issues included improved privacy settings, content moderation, and digital literacy initiatives to help users navigate the online landscape effectively.

The mobile internet, with the advent of smartphones, propelled communication into the palm of our hands. Now, we could access the internet on the go, making us constantly connected and reachable. This has transformed our social dynamics, blurring the boundaries between work and personal life, as well as reshaping the way we consume news, entertainment, and information.

The proliferation of mobile applications, or apps, allowed us to perform tasks ranging from ordering food and conducting financial transactions to navigating directions and tracking our health. These apps, accessible through app stores, have provided immense convenience and have propelled the growth of the app economy, creating new opportunities for innovation and entrepreneurship.

Virtual reality (VR) has emerged as another groundbreaking communication technology. By immersing us in computer-generated environments, VR creates a sense of presence and interaction like never before. The technology has come a long way since its inception, with modern VR headsets offering high-resolution displays, precise motion-tracking, and realistic haptic feedback.

VR has found applications in various industries, including gaming, entertainment, education, healthcare, and even professional training. With VR, we can now visit distant destinations, explore simulated historical worlds, practise complex surgical procedures, or walk on the surface of Mars – all from the comfort of our homes or offices.

Moreover, the combination of VR and the internet has given birth to social virtual reality, where people can interact with each other in shared virtual spaces. This opens up new avenues for socialising, telecommuting, and experiencing events and environments that were previously inaccessible. Collaborative virtual reality also holds tremendous potential for remote teams to work together, fostering collaboration and creativity.

However, as with any technological advancement, there are challenges and ethical considerations associated with these communication breakthroughs. The internet has given rise to concerns about online privacy, cybersecurity, and the ethical use of personal data. Governments and organisations have had to establish regulations to protect individuals' rights while enabling the global flow of information.

Similarly, virtual reality raises questions

about the potential impact on mental health, addiction, and the boundaries between reality and simulation. Researchers and developers continue to explore ways to mitigate potential adverse effects, ensure inclusive and accessible experiences, and push the boundaries of VR innovation.

As communication technology continues to evolve, it is essential to address these challenges and ensure that these breakthroughs are used responsibly and ethically. By leveraging the power of communication technology effectively, we can create a more connected world, where information flows freely, and individuals are empowered to collaborate, learn, and grow, ultimately driving positive societal change.

In the following chapters, we will explore how communication breakthroughs have impacted various sectors, such as medicine, energy, transportation, and space exploration. These advancements are interconnected, forming a tapestry of progress driven by the convergence of science and technology. The potential for even further transformations in communication technology is vast, and it is an exciting time to witness and participate in this incredible journey.

CHAPTER FIVE

Medicine and Health: Revolutionising Healthcare

The field of medicine and healthcare has undergone remarkable transformations in recent years, propelled by advancements in science and technology. These groundbreaking achievements have revolutionised the way we diagnose, treat, and prevent illnesses, ultimately leading to improved health outcomes and enhancing the quality of life for people all around the world.

One of the most impactful advancements in healthcare is the emergence of precision medicine. This innovative approach takes into account an individual's unique genetic make-up, lifestyle, and environmental factors to tailor treatment plans specifically for them. By analysing an individual's genetic profile, doctors can identify specific genetic mutations or variations that predispose them to certain diseases, allowing for more targeted interventions and personalised treatment strategies. Traditionally, the "one-size-fits-all" approach to medicine often resulted in suboptimal outcomes, as individuals with the same disease might respond differently to a particular treatment due to subtle genetic variations. Therefore, precision medicine has not only revolutionised patient care but is also minimising adverse drug reactions and unnecessary medical interventions.

In addition to precision medicine, the integration of digital health technologies has also played a pivotal role in transforming the healthcare landscape. The advent of wearable devices, mobile apps, and telemedicine platforms has empowered individuals to actively monitor their health and communicate with healthcare professionals remotely. Wearable devices, such as fitness trackers and smartwatches, provide real-time data on physical activity, heart rate, sleep patterns, and more, enabling individuals to proactively manage their health and make informed decisions regarding their lifestyle choices. Moreover, these devices can help detect early warning signs of various health conditions, allowing for timely intervention and prevention. Telemedicine platforms have also gained significant traction, particularly during the COVID-19 pandemic, as they allow patients to consult with healthcare providers through video calls or online chats. This not only reduces the need for physical visits but also enhances accessibility to medical care, particularly in remote or underserved areas. The convenience and efficiency of digital health technologies have led to improved patient engagement, reduced healthcare costs, and better health outcomes.

Furthermore, the utilisation of big data and artificial intelligence (AI) in healthcare has had a profound impact on disease understanding, prevention, and management. The vast amount of healthcare data generated from electronic health records, clinical trials, research studies, and patient monitoring systems can now be analysed using sophisticated algorithms and machine learning techniques. These technologies enable healthcare providers to identify trends, predict disease outbreaks, and develop targeted interventions. AI-powered systems can assist in diagnosing diseases with remarkable accuracy and speed, leading to early detection and intervention. For instance, deep learning algorithms can analyse medical images, such as X-rays and MRIs, to detect abnormalities that may be missed by human observers. By leveraging AI and big data, healthcare professionals can make more informed decisions, improve patient outcomes, and enhance population health management. Moreover, AI algorithms can analyse vast amounts of medical literature and research data, enabling rapid and comprehensive literature reviews, which are critical for evidence-based medicine and advancing medical knowledge.

The evolution of medical imaging technologies has also been instrumental in reshaping healthcare practises. Techniques such as magnetic resonance imaging (MRI), computed tomography (CT), and positron emission tomography (PET) have significantly enhanced our ability to visualise and analyse the internal structures of the body. These high-resolution imaging modalities enable more accurate diagnoses, precise surgical planning, and targeted therapeutic interventions. Integrated imaging modalities, such as PET-CT and PET-MRI, have revolutionised the evaluation of diseases like cancer and neurodegenerative disorders, providing both anatomical and functional information in a single examination. Furthermore, advancements in imaging technology have led to the development of molecular imaging, which utilises radiotracers to visualise and analyse biological processes at the cellular and molecular level, aiding in early disease detection and personalised treatment selection.

The field of regenerative medicine is yet another frontier in healthcare revolution. Stem cell research, tissue engineering, and gene therapy hold the promise of restoring and repairing damaged tissues and organs. Stem cells, with their remarkable ability to differentiate into various cell types, offer the potential to regenerate tissue in conditions such as spinal cord injuries, heart disease, and degenerative neurological disorders. Tissue engineering approaches involve combining cells, biomaterials, and biochemical cues to create functional tissues and organs in the laboratory for transplantation. Gene therapy aims to correct genetic mutations or introduce therapeutic genes to improve cell and tissue function. While regenerative medicine is still in its early stages, ongoing research and clinical trials show promising results, and it is anticipated to have a profound impact on the treatment of a wide range of diseases and medical conditions. However, challenges related to safety, efficacy, and the ethical use of stem cells need to be addressed before these therapies become routine clinical practises.

While these remarkable advancements have the potential to revolutionise healthcare, they also give rise to several ethical and societal considerations. The collection and use of personal health data raise concerns about privacy and security. Safeguarding patient confidentiality and ensuring responsible use of these data are essential for building trust in the healthcare system. Additionally, questions regarding the accessibility and affordability of these technologies need to be addressed to ensure equitable healthcare for all. Disparities in access to digital health tools, genetic testing, and specialised healthcare services can exacerbate existing health inequities. It is imperative for policymakers, healthcare professionals, and society as a whole to carefully navigate these challenges and ensure that the benefits of these advancements are equitably distributed while safeguarding patient autonomy, privacy, and dignity.

In conclusion, the field of medicine and healthcare is experiencing an unprecedented revolution, driven by advancements in science and technology. Precision medicine, digital health technologies, artificial intelligence, medical imaging, and regenerative medicine are all reshaping the way we approach and deliver healthcare. By embracing these advancements responsibly, we have the potential to improve health outcomes, redefine the patient experience, and transform the future of healthcare. However, addressing ethical, legal, and social implications is crucial to ensure that these innovations are deployed in an equitable and ethical manner. By working collectively to overcome these challenges, we can unlock the full potential of these advancements and create a healthier future for all.

CHAPTER SIX

Energy and Sustainability in the 21st Century

I n the 21st century, the pressing need for energy production and the urgency to address climate change have set the stage for a monumental shift in how we approach energy sustainability. As the global population continues to grow, the demand for energy is expected to increase significantly, posing significant challenges to traditional energy systems that heavily rely on fossil fuels. In this chapter, we will dive deeper into the current state of energy production and explore the multitude of efforts being made towards achieving sustainability.

One of the primary concerns associated with traditional energy sources, such as coal and oil, is their environmental impact. Combustion of these fuels releases a substantial amount of greenhouse gases into the atmosphere, contributing to global warming and climate change. Additionally, the extraction and processing of fossil fuels often lead to habitat destruction and water pollution. It is evident that a transition towards cleaner and more sustainable energy sources is imperative to secure a healthier planet for future generations.

Renewable energy sources, including solar, wind, hydro, and geothermal power, have gained significant momentum in recent years as viable alternatives to fossil fuels. Solar power utilises photovoltaic cells to convert sunlight into electricity, while wind power harnesses the kinetic energy of moving air to generate electricity. Hydroelectric power generates electricity by utilising the potential energy of stored water in reservoirs, whereas geothermal power harnesses the heat from the Earth's core. Each of these sources offers unique advantages and has the potential to significantly reduce greenhouse gas emissions.

Solar energy, in particular, has witnessed a remarkable proliferation due to technological advancements and declining costs. Photovoltaic cells are now more efficient at converting sunlight into electricity, and the market for solar panels has experienced rapid growth. In addition, large-scale solar power plants and rooftop solar installations have become commonplace, allowing individuals and businesses to generate their electricity and reduce their reliance on fossil fuels.

The wind energy sector has also experienced

remarkable growth, with wind turbines becoming larger, more efficient, and more cost-effective. Offshore wind farms have emerged as a promising solution due to stronger and more consistent wind speeds. The availability of wind resources in coastal areas and offshore locations allows for the deployment of larger wind turbine arrays, maximising energy generation potential.

Hydropower, a well-established source of renewable energy, continues to play a crucial role in meeting energy demands while producing zero emissions. Hydroelectric power plants generate electricity by converting the kinetic energy of flowing or falling water into mechanical energy, which is then used to drive turbines and generators. However, the construction of large dams for hydroelectric projects may have adverse environmental and social impacts, necessitating careful planning and mitigation measures.

Geothermal energy, although less commonly utilised, has significant potential, especially in areas with high geothermal activity. Geothermal power plants tap into the Earth's natural heat reservoirs to generate electricity or directly provide heating and cooling for buildings. The continuous access to geothermal energy offers a reliable and baseload source of power, reducing the need for intermittent sources like solar and wind.

To overcome the challenges associated with the intermittent nature of renewable energy sources, energy storage systems have become essential for maintaining a stable and reliable energy supply. Batteries, flywheels, and other storage technologies have made significant advancements, allowing excess energy to be stored and used during periods of low renewable energy production or high demand. These storage technologies facilitate the integration of renewable energy into existing grids, minimising reliance on conventional backup power sources.

In addition to transitioning towards renewable energy sources, improving energy efficiency has become a significant focus in the pursuit of sustainability. Energy efficiency measures aim to reduce overall energy consumption while maintaining or enhancing productivity. Advanced technologies, such as smart grids and energy management systems, enable better monitoring and control of energy usage, leading to optimised energy consumption.

Buildings, which account for a significant portion of global energy consumption, have become a target for energy efficiency measures. Innovative building designs, improved insulation, and energy-efficient appliances have collectively contributed to significant reductions in energy consumption. Moreover, the adoption of energy-efficient lighting, such as LEDs, has not only reduced energy demand but also increased the lifespan of lighting systems.

Transportation, another major energy consumer, is undergoing a revolution in terms of sustainability. Electric vehicles (EVs), powered by electricity rather than fossil fuels, have gained popularity due to advancements in battery technology and increased charging infrastructure. The adoption of EVs not only reduces greenhouse gas emissions but also promotes energy diversification by utilising renewable energy sources. Additionally, initiatives such as car-sharing services and improved public transportation systems encourage a shift away from private vehicles, further reducing energy consumption and emissions.

Governments worldwide have recognised the need to prioritise sustainability in their energy plans. A considerable number of countries have set ambitious targets to reduce greenhouse gas emissions and increase the share of renewable energy in their energy mix. To encourage the adoption of sustainable energy technologies, governments have implemented various policy mechanisms, including feed-in tariffs, tax incentives, and renewable portfolio standards. These policies aim to create a favourable environment for investment in renewable energy and foster the development of a self-sustaining, clean energy industry.

International collaborations and agreements also play a crucial role in achieving sustainability goals. The Paris Agreement, signed by nearly all countries, sets out a roadmap to limit global temperature rise and strengthen the global response to climate change. Countries commit to regular reporting, reassessment, and scaling up their efforts to achieve a low-carbon economy. Through cooperation and knowledge-sharing, countries can collectively work towards a more sustainable energy future.

While the transition to a sustainable energy future holds immense promise, it does come with challenges. The costs associated with de-

veloping and implementing renewable energy technologies can be significant, requiring
substantial investment and financial resources.
Policies should be in place to incentivise renewable energy deployment and ensure a level playing field for all energy sources. The reliance on traditional fossil fuel industries for
employment and economic growth presents a
challenge in transitioning away from these industries without causing social and economic upheaval. Therefore, transitioning to sustainable energy must consider the implications
for communities and workers affected by this
transformation and explore opportunities for
a just transition.

The energy sector is also grappling with issues surrounding grid integration and stability. Renewable energy sources, such as solar
and wind, are inherently intermittent, mean-
ing that their output fluctuates depending on
variables like weather conditions. This inter-
mittency poses challenges to grid operators,

who are responsible for maintaining a reliable supply of electricity. To address this, innovative solutions like demand-response programmes and virtual power plants are being explored. These programmes enable grid operators to manage electricity demand in real-time and leverage distributed energy resources like residential solar panels and battery storage to ensure grid stability.

Additionally, technology development and innovation play crucial roles in driving the transition to a sustainable energy future. Researchers and scientists around the world are continuously working towards improving the efficiency, reliability, and affordability of renewable energy technologies. Advancements in materials science, such as the development of more efficient solar cell materials or advanced battery technologies, have the potential to revolutionise the energy landscape. Furthermore, emerging concepts like artificial photosynthesis, where sunlight is directly used to produce fuels, offer exciting prospects for overcoming the limitations of renewable energy generation.

It is essential to recognise that achieving sustainability in the energy sector is not a solitary endeavour. Collaboration between governments, businesses, communities, and individuals is vital for success. Local communities and grassroots organisations play a crucial role in advocating for renewable energy projects and influencing local energy policies. Businesses also have a responsibility to embrace sustainable practises by setting ambitious sustainability goals, investing in renewable energy technologies, and adopting energy-efficient practises.

Individuals can contribute to sustainability by making conscious choices in their energy consumption, such as reducing energy waste, using energy-efficient appliances, and supporting renewable energy initiatives. Additionally, individuals can advocate for sustainable energy policies and educate others about the importance of transitioning to renewable energy sources.

Overall, the transition to a sustainable energy future requires a multidimensional approach that encompasses policy changes, technological advancements, and a shift in societal attitudes towards energy consumption. It is a complex and challenging task, but one that is necessary to mitigate the impacts of climate change and ensure a sustainable future for generations to come. By working together and embracing innovative solutions, we can create a cleaner and more sustainable energy system that supports both human well-being and the health of our planet.

CHAPTER SEVEN

Space Exploration and the Colonisation of Other Planets

H uman beings have always looked up at the night sky with a sense of wonder and curiosity. Over the centuries, we have made incredible advancements in our understanding of the universe and our ability to explore it. Space exploration has become not only a scientific endeavour but also a symbol of human ingenuity and our desire to push the boundaries of what is possible.

In recent years, the idea of colonising other planets has gained increasing attention. With the rapid depletion of resources on Earth and the growing concerns about our planet's sustainability, the possibility of establishing colonies on other celestial bodies has become an enticing prospect. It represents our collective ambition to explore, expand, and evolve.

One of the primary candidates for colonisation is Mars, often referred to as the "Red Planet." Mars shares several characteristics with Earth, such as the presence of water in the form of ice, a carbon dioxide atmosphere, and a similar day length. These similarities make Mars a potentially habitable planet for human beings.

NASA, along with other space agencies and private companies, has been actively researching and planning missions to Mars. The goal is not only to explore the planet but also to prepare for future human colonisation. Robotic missions have already provided valuable data about Mars' environment, geology, and potential resources. These missions have also enabled us to test technologies and systems that will be essential for human survival on the planet.

But colonising Mars is not without its challenges. The journey alone would take several months, requiring astronauts to endure long periods of isolation, confinement, and exposure to radiation. Once on Mars, settlers would have to deal with extreme temperatures, a thin atmosphere, and the absence of a readily available food and water source. The development of sustainable habitats, reliable life support systems, and efficient resource utilisation would be critical for long-term colonisation.

To tackle these challenges, scientists and engineers are exploring innovative solutions, such as 3D printing technologies to construct habitats using local resources, in-situ resource utilisation to extract water and produce oxygen, and advanced propulsion systems to shorten travel times. These advancements are essential in making permanent human presence on Mars a reality.

Beyond Mars, other candidates for colonisation include the moon, which would serve as a stepping stone for further space exploration, and potentially, other moons and planets within our own solar system. The idea of establishing self-sustaining colonies in these locations raises profound questions about the future of human civilisation and our potential as an interplanetary species.

Space exploration and colonisation offer not only the potential for new scientific discoveries but also the opportunity for humanity to ensure its long-term survival. By expanding our reach beyond Earth, we can reduce the risks associated with overpopulation, resource depletion, and natural disasters. Additionally, the establishment of colonies on other planets would provide a platform for scientific research, innovation, and the exploration of the unknown.

However, the pursuit of space exploration and colonisation must be carried out with great caution and responsibility. We must consider the ethical implications of our actions, including the preservation of extra-terrestrial environments and the potential impact on indigenous life forms, if they exist. It is crucial to approach colonisation efforts with a mindset of sustainability, respect for the universe's inherent value, and a commitment to preserving the delicate balance of life.

As we venture further into space, we must also remember the importance of collaboration. International cooperation will be vital in realising the dream of interplanetary colonisation. By pooling our resources, knowledge, and expertise, we can overcome the challenges that lie ahead and ensure a future where human beings inhabit not only Earth but other celestial bodies as well.

The colonisation of other planets is not merely a matter of scientific curiosity; it is an investment in our future as a species. It represents our collective ambition to explore, expand, and evolve. As we look to the stars and imagine the possibilities, let us remember to approach this endeavour with humility, curiosity, and a deep appreciation for the wonders of the universe.

Furthermore, colonisation efforts will undoubtedly have significant implications for society and governance. Establishing new colonies on other planets raises questions about jurisdiction, legal frameworks, and the rights of those who choose to leave Earth. Policies and agreements will need to be developed to ensure fairness, equality, and cooperation among all parties involved.

Moreover, the colonisation of other planets has the potential to revolutionise various scientific disciplines. The study of extraterrestrial life, for instance, could provide invaluable insights into the origin and definition of life itself. It could challenge our understanding of biochemistry, genetics, and evolutionary biology. The presence or absence of life on other planets would fundamentally alter our perception of the universe and our place within it.

In addition to scientific advancements, the colonisation of other planets could lead to the development of new technologies and industries. Mining asteroids for valuable resources, such as rare metals or water, could allow us to address resource scarcity on Earth and open up new avenues for economic growth. The challenges encountered in establishing habitats and sustaining life in extreme environments would drive innovation in areas such as renewable energy, agriculture, and waste management.

Ultimately, the exploration and colonisation of other planets may also provide a sense of unity and purpose for humanity. As we venture beyond the confines of Earth and face the immense challenges of space travel and settlement, we may find ourselves transcending the divides that often plague our world. Collaboration on a global scale could foster a sense of shared destiny and responsibility for the future of our civilisation.

Furthermore, the colonisation of other planets offers the potential for cultural enrichment and the preservation of human knowledge. As civilisations expand beyond Earth, diverse perspectives, traditions, and art forms will merge, creating a melting pot of ideas and creativity. The exchange of cultural practises could inspire new forms of expression and bring people closer together, emphasising our common humanity even across vast distances.

In conclusion, the exploration and colonisation of other planets hold the potential to shape the future of human civilisation. Beyond being a scientific endeavour, it represents a bold endeavour to ensure the survival, progress, and collective evolution of our species. As we embark on this journey, may we do so with a deep sense of respect for the universe, an unwavering commitment to sustainability and ethical responsibility, and a profound appreciation for the remarkable possibilities that lie beyond our home planet.

CHAPTER EIGHT

The Future of Transportation: From Flying Cars to Hyperloop

The future of transportation holds exciting possibilities as advancements in technology continue to reshape the way we move from one place to another. In this chapter, we delve deeper into some of the most promising modes of transportation that could become a reality in the coming decades, exploring their potential benefits, challenges, and implications.

One concept that has captured the imaginations of both scientists and enthusiasts is the idea of flying cars. While flying cars have long been a staple of science fiction, recent developments in electric propulsion and autonomous technology have brought us closer to turning this dream into a reality. Companies like Uber and Airbus are already testing prototypes of vertical takeoff and landing (VTOL) vehicles, which could potentially revolutionise urban mobility by combining the advantages of helicopters and cars. Imagine being able to bypass traffic congestion and soar above the cityscape, reaching your destination in a fraction of the time it would take by conventional means.

The advantages of flying cars extend beyond just convenience and speed. They could potentially have a significant positive impact on urban planning and infrastructure. With the ability to both take off and land vertically, flying cars would eliminate the need for extensive road networks and parking spaces, freeing up valuable land for other purposes. Additionally, flying cars could provide a solution to the last-mile problem in transportation, facilitating efficient and direct travel to even the most remote areas.

However, the widespread adoption of flying cars also presents significant challenges. One major hurdle is the development of safe and reliable autonomous technology to navigate these vehicles. Ensuring that flying cars can operate without human error or interference is crucial to their success. Autonomous technology must be able to handle complex scenarios such as air traffic coordination, emergency situations, and adverse weather conditions. Furthermore, the integration of flying cars into existing airspace systems needs careful consideration to avoid congestion, ensure safety, and address privacy concerns.

In terms of infrastructure, building the necessary infrastructure to support a fleet of flying cars poses substantial challenges. Vertiports—specialised landing pads and takeoff points—would need to be strategically located throughout urban areas to promote efficient travel routes. These vertiports would require advanced facilities for charging, maintenance, and passenger services. Urban planning and regulatory frameworks would need to be adapted to accommodate this new mode of transportation and ensure its safe integration into the existing urban fabric.

Another transportation innovation that shows great promise is the Hyperloop. Conceived by visionary entrepreneur Elon Musk, the Hyperloop is a high-speed transportation system that uses low-pressure tubes to propel pods at incredible speeds. This concept has the potential to revolutionise long-distance travel, with speeds that have been projected to reach up to 700 mph (1126 km/h). Not only would this dramatically reduce travel times between major cities, but it would also have a substantially smaller environmental impact compared to traditional modes of transport, such as aeroplanes or vehicles.

The Hyperloop, with its potential to connect distant cities and regions seamlessly, could have far-reaching implications for urban development, tourism, and economic growth. It would bridge the gap between regions, making them more accessible and interconnected. This could lead to the creation of new job opportunities, the revitalisation of rural areas, and the redistribution of resources across regions.

However, the development and implementation of the Hyperloop pose significant challenges. The technical engineering requirements, such as designing a low-pressure tube system that allows for high-speed travel while maintaining passenger safety and comfort, are complex. Ensuring adequate ventilation, minimising noise pollution, and addressing potential maintenance and durability issues are key considerations. Additionally, securing right-of-way for the Hyperloop's infrastructure and obtaining necessary regulatory approvals pose significant hurdles. The sheer scale of implementation and the cost involved also need careful consideration.

In addition to flying cars and the Hyperloop, other advancements in transportation technology are also worth mentioning. Electric vehicles (EVs) are becoming increasingly popular as improvements in battery technology allow for longer ranges and faster charging times. The widespread adoption of EVs would not only reduce greenhouse gas emissions but also decrease our reliance on fossil fuels. Governments and industries around the world are investing heavily in the development of charging infrastructure and incentives to promote electric vehicle adoption.

Autonomous vehicles (AVs) are another significant development in transportation technology. With AI-powered systems capable of navigating and driving with minimal human intervention, the potential for safer and more efficient roads is within reach. AVs have the potential to reduce traffic congestion, optimise fuel consumption, and improve traffic flow through advanced communication and coordination systems. Additionally, they could enhance accessibility for people with disabilities and older adults, enabling independent mobility and social participation.

However, the adoption of autonomous vehicles raises complex questions regarding ethics, liability, and the future of employment. As self-driving cars become the norm, how will responsibility be assigned in case of accidents? Will there be a need for a new regulatory framework to ensure safety and compliance? Moreover, the widespread implementation of AVs may result in job displacement in the transportation industry, particularly for professional drivers. Governments and industries must proactively address these concerns to ensure a smooth transition and offer support to affected workers.

As we look ahead to the future of transportation, it is important to acknowledge the challenges and considerations that come with these innovations. Regulatory, safety, and infrastructure concerns must be addressed to ensure the successful integration of these technologies into our daily lives. Ethical questions, such as how to balance the benefits of technology with concerns over privacy and job displacement, also need to be carefully considered.

In conclusion, the future of transportation holds tremendous potential for transformative change. From flying cars to the Hyperloop, advancements in technology are paving the way for faster, more efficient, and sustainable modes of transportation. As we embrace these innovations, it is crucial to approach their development and implementation with careful consideration and a focus on creating a future of transportation that is safe, accessible, and environmentally conscious.

CHAPTER NINE

Exploring the Uncharted

The human desire to explore the unknown has always been a driving force throughout history. From the great voyages of discovery to the conquest of space, humans have continuously pushed the boundaries of what is known and ventured into uncharted territories. In this captivating chapter, we delve into the fascinating world of exploration and the endless possibilities that lie ahead, unravelling the depths of our curiosity and the unrelenting spirit of adventure.

1. The Age of Exploration:

The Age of Exploration marked a pivotal moment in history when brave explorers set sail to discover new lands and trade routes. From Christopher Columbus to Ferdinand Magellan, these individuals undertook perilous journeys across oceans, facing unknown dangers and hardships. Their quest extended the boundaries of human knowledge, opened up new trade routes, brought about cultural exchanges, and forever changed our understanding of the world. These expeditions played a crucial role in shaping the development of civilisation, introducing new goods, ideas, and cultures that transformed societies across the globe.

2. Unearthing the Mysteries of the Ocean:

Despite the vast majority of the Earth being

covered in water, we have only scratched the surface of understanding what lies beneath the ocean's depths. The exploration of the oceans has revealed breathtaking biodiversity and hidden wonders, such as the Great Barrier Reef and the Mariana Trench. With new technologies, such as deep-sea submarines and underwater drones, scientists are venturing deeper into the unknown, uncovering new species, undiscovered ecosystems, and unlocking the secrets of our planet's past.

Ocean exploration is not just about discovering new species; it also plays a vital role in understanding climate change. By studying the ocean's currents, marine life, and geological formations, scientists can gain crucial insights into the impact of human activity on the environment. Furthermore, the exploration of the ocean's depths has led to technological advancements in underwater robotics, providing us with the tools to examine and protect fragile ecosystems and even search for remnants of ancient civilisations submerged beneath the waves.

3. Mapping the Cosmos:

The exploration of space has captivated the human imagination for centuries. From the early observations of the stars and planets to the remarkable achievements of space agencies like NASA and ESA, we have gained a remarkable understanding of our place in the universe. Advances in astronomy and astrophysics have allowed us to explore distant galaxies, uncover the mysteries of black holes, and search for signs of extra-terrestrial life. The quest to comprehend the vastness of the cosmos continues to push the boundaries of human knowledge.

As technology evolves, so does our ability to explore the cosmos. Telescopes, such as the Hubble Space Telescope and the forthcoming

James Webb Space Telescope, enable us to peer deeper into the universe, capturing breathtaking images and unravelling the secrets of its creation. The emergence of space tourism and private space companies also promises to revolutionise space exploration, potentially opening up opportunities for civilians to venture beyond Earth's atmosphere and witness the vast beauty of the cosmos first-hand.

4. Uncharted Frontiers: The Final Frontier:

As we continue to expand our exploration of space, our sights have now turned towards the colonisation of other planets. Mars, with its similarities to Earth, has become the focus of much attention and research. Private space companies, along with government agencies, are investing in developing technologies to enable human colonisation of Mars, envisioning a future where interplanetary travel and settlement become a reality. The exploration of Mars and other celestial bodies will not only expand our horizons but also bring about new scientific discoveries and advancements in technology that will benefit humanity as a whole.

Additionally, the pursuit of space exploration has numerous practical applications for life on Earth. Technologies developed for space travel, such as solar energy, water purification systems, and lightweight materials, have already found their way into various industries, improving our daily lives and fostering sustainable practises. Furthermore, space exploration promotes international collaboration as countries around the world join forces to tackle the challenges and reap the rewards of venturing into the final frontier.

Conclusion:

Exploring the uncharted is a testament to the human spirit and our innate curiosity. It is through exploration that we push the limits of our knowledge, discover new possibilities, and redefine what is possible. Each step we take toward understanding the oceans and unravelling the secrets of the cosmos brings us closer to unlocking the mysteries of our existence. As we venture into the unknown, let us embrace the challenges, pursue scientific breakthroughs, and continue to shape a future filled with endless opportunities for exploration and discovery.

CHAPTER TEN

The Power of Imagination in Shaping the Future

I n the vast realm of human achievement, the
power of imagination stands as an unparalleled force driving progress and shaping the
future. It is through the fertile landscape of our
minds that visionary ideas are born, creativity is sparked, and innovative breakthroughs
take flight. This chapter delves into the profound influence of imagination on the trajec-
tory of human development, exploring how it

has propelled us forward in countless domains, and how harnessing its potential can pave the way for a bright and prosperous future.

Unlocking Boundless Potential:

Imagination serves as a limitless reservoir of ideas, enabling us to visualise and conceptualise what has yet to come into existence. It propels us beyond the confines of our current limitations, empowering us to dream big and aim high. Throughout history, it has been the driving force behind revolutionary inventions, artistic masterpieces, and scientific discoveries.

Consider the awe-inspiring works of Jules Verne, whose vivid imagination gave birth to novels like "Twenty Thousand Leagues Under the Sea" and "From the Earth to the Moon." Verne's writings painted a picture of a future where submarines explored the deep ocean and humans ventured into space. Such visions of the future, fuelled by imagination, inspired generations to pursue scientific advancements that turned Verne's fantasies into reality.

In science, imagination plays a crucial role in hypothesis formation and experimentation. Einstein's theory of relativity, for instance, was born from his thought experiments and imaginative leaps. By allowing his mind to surpass conventional thinking, he was able to redefine our understanding of space, time, and gravity. Imagination sparked his curiosity and propelled him towards uncovering the mysteries of the universe.

Inspiring Innovation and Progress:

Imagination acts as a catalyst for innovation, igniting the spark that fuels progress and propels societies forward. It encourages us to question the status quo, challenge ingrained beliefs, and seek novel solutions to complex problems. By envisioning a future free from the constraints of convention, we cultivate a mindset that encourages experimentation, risk-taking, and the pursuit of disruptive ideas.

Consider the case of Steve Jobs, whose imaginative vision revolutionised the technology industry. Jobs, driven by his innate ability to think differently, imagined a world where computers were intuitive and accessible to all. This vision led to the creation of game-changing products such as the Macintosh, iPod, and iPhone, transforming multiple industries and shaping the way we connect, communicate, and consume information.

In addition to technological advancements, imagination has also played a pivotal role in shaping artistic expressions that challenge societal norms. Often, artists push the boundaries of traditional aesthetics and conventions, presenting new perspectives and provoking thought. From the artistic movements of sur-realism to avant-garde theatre performances, imagination opens the door to innovative and boundary-breaking creative expressions that push us to reconsider our preconceptions.

Imagination and Social Change:

Beyond its impact on technological and artistic advancements, imagination is a pow-erful tool for bringing about social change. It allows us to empathise with the experi-ences of others, envision a more equitable so-ciety, and mobilise efforts towards achieving it. Throughout history, visionary leaders and activists have harnessed the power of imagina-tion to challenge societal norms, advocate for justice, and foster inclusivity.

Consider the pioneering work of Martin

Luther King Jr., whose powerful speeches painted a vivid picture of a future free from racial discrimination and inequality. His dream, powerfully conveyed through words and imagination, galvanised a nation to fight for civil rights and set in motion a monumental shift in societal attitudes and policies. Imagination serves as a vehicle for social imagination, allowing us to transcend our own experiences and envision a more just and compassionate society.

Nurturing Imagination in Education:

Recognising the intrinsic value of imagination, educational institutions are increasingly emphasising the cultivation of this vital skill. Educators are incorporating creative thinking and problem-solving exercises into curricula, providing students with the tools to unleash their imagination. By fostering an environment that encourages curiosity, exploration, and innovation, we equip future generations with the ability to tackle the challenges of tomorrow and shape a world that reflects their visionary ideals.

One standout example is Finland's education system, which prioritises play, creativity, and imagination. Finnish schools emphasise collaborative, inquiry-based learning, allowing students to explore their interests and develop a sense of wonder. This approach fosters not only innovation but also resilience and adaptability, skills crucial for navigating an ever-changing landscape in the future.

The Responsibility of Imagination:

While the power of imagination has immense potential for positive change, it also comes with great responsibility. As we explore new frontiers and push the boundaries of what is possible, ethical considerations must guide our imaginative endeavours. We must tread carefully to ensure that imagination is harnessed for the greater good, respecting the boundaries of morality, sustainability, and the well-being of individuals and communities.

For instance, in the field of artificial intelligence, creating and imagining intelligent systems requires strict ethical guidelines to protect human dignity and prevent harm. Ethical discussions around privacy, bias, and accountability are essential to ensure that imagination-driven technological advancements align with our values and aspirations.

Furthermore, the power of imagination should not be limited to a select few but must be made accessible to all individuals, regardless of socioeconomic status or background. By fostering inclusivity and providing equal opportunities for imaginative exploration, we create a society where diverse perspectives and experiences contribute to a richer tapestry of collective imagination.

Conclusion:

As we reflect on the power of imagination in shaping the future, we are reminded of its transformative influence across all spheres of human endeavour. It is through creative leaps of imagination that we continue to surmount seemingly insurmountable challenges, unlocking the potential of a future yet to be realised. By embracing this extraordinary faculty of the human mind and channelling it towards honourable pursuits, we hold the key to a future that surpasses our wildest dreams. Let our imagination soar, for it is through its power that we shall shape a world beyond our current imaginings.

CHAPTER ELEVEN

Unveiling the Unknown

In the vast realm of science and technology, there lies an enigmatic and tantalising frontier that has captivated human curiosity for centuries: the unknown. Throughout history, humanity has embarked on a relentless pursuit of knowledge, exploring the mysteries of the universe and pushing the boundaries of our understanding. In this chapter, we embark on a journey through the uncharted territories of science and technology as we unveil the unknown and delve into the realm of the unexplored.

The Quest for Knowledge

From the ancient philosophers pondering the nature of existence to the modern-day scientists unravelling the complexities of the cosmos, the desire to uncover the unknown has been a driving force in human civilisation. The quest for knowledge has led us to make ground-breaking discoveries, challenging our preconceived notions and revolutionising our understanding of the world.

Unravelling the Mysteries of the Universe

One of the most fascinating domains where we strive to unveil the unknown lies in our exploration of the universe. Through telescopes, space probes, and sophisticated instruments, we have embarked on a cosmic journey, seeking answers to questions that have perplexed us for centuries. The birth of the universe, the formation of galaxies, the enigma of dark matter and dark energy, the nature of black holes—these cosmic mysteries continue to beckon us, enticing us to probe deeper and uncover the secrets that lie beyond our reach.

Recent advancements in observational technology, such as gravitational wave detection, have opened up new avenues of exploration. We learn previously unheard-of things about the nature of the universe by observing and researching the ripples in spacetime that catastrophic cosmic events cause. These discoveries are like windows into the unknown, revealing the hidden intricacies of celestial phenomena and expanding our understanding of the cosmos.

In addition to our quest to understand the broader universe, we are also exploring the mysteries found within our very own solar system. Missions to distant planets, moons, and asteroids offer glimpses into the geological and atmospheric wonders beyond Earth. The exploration of Mars, with its potential for past or present microbial life, has fuelled our curiosity and propelled us closer to understanding the potential for life beyond our planet.

Unearthing the Secrets of the Mind

While the vastness of the cosmos may seem infinite, a different frontier of exploration lies within ourselves—the human mind. For centuries, psychologists and neuroscientists have endeavoured to unravel the complexities of our thoughts, emotions, and consciousness. Through ground-breaking research and advancements in technologies such as brain imaging and artificial intelligence, we are beginning to lift the veil on the inner workings of the brain, unravelling the mysteries of human cognition and paving the way for new frontiers in understanding consciousness itself.

Neuroscience has revealed the brain's remarkable plasticity and its ability to adapt and rewire itself throughout life. This new knowledge could lead to new ways to treat neurological disorders and improve cognitive abilities, but it also brings up important ethical questions about identity, privacy, and what might happen if we change the most basic parts of what it means to be human.

The emerging field of connectomics aims to map the intricate neural networks of the brain, offering insights into how our thoughts, memories, and emotions are encoded within its complex web of connections. This field has the potential to unlock revolutionary treatments for mental illnesses, enhance our understanding of consciousness, and give us unprecedented control over our own minds. However, as we explore these uncharted territories, we must navigate the ethical dimensions of cognitive enhancement, artificial intelligence, and privacy, ensuring that our advances in the understanding of the mind are used responsibly and for the betterment of humanity.

Emerging Technologies and the Unknown

As our understanding of science and technology advances, so too does the realm of the unknown. Emerging technologies such as quantum computing, nanotechnology, and genetic engineering hold the promise of unlocking new frontiers in our understanding of the universe and the human experience. Quantum computing, with its potential to tackle complex problems that classical computers cannot solve, could revolutionise fields such as cryptography, drug discovery, and optimisation algorithms. Nanotechnology, the science of manipulating matter at the atomic and molecular scale, could lead to innovative materials, enhanced medical treatments, and advancements in energy production. Genetic engineering, with the ability to edit the genome, has the potential to eradicate genetic diseases, improve crop yields, and even reshape the evolutionary path of life on Earth.

The unknown territories of these emerging technologies are not without challenges. Quantum computing, for example, grapples with the peculiarities of quantum mechanics, where particles can exist in multiple states simultaneously. The control and stability of quantum systems present significant obstacles, requiring new breakthroughs in materials science and quantum error correction. Nanotechnology faces ethical dilemmas surrounding the responsible use of nanomaterials, their potential environmental impact, and unforeseen consequences of manipulating matter at such a small scale. Genetic engineering raises concerns about unintended consequences, genetic discrimination, and the ethical boundaries of interventions on the human germline.

Embracing the Unknown

In the quest to unveil the unknown, we must approach these uncharted territories with caution, humility, and curiosity. We must recognise that the unknown represents both possibilities and challenges, and that every discovery brings with it new questions and responsibilities. As we venture deeper into the unexplored realms of science and technology, let us embrace the unknown not with fear, but with a sense of wonder and reverence for the mysteries that lie ahead.

Our journey into the unknown draws us closer to unravelling the cosmic mysteries, understanding the complexities of the mind, and harnessing the potential of emerging technologies. It is through our collective effort, curiosity, and relentless pursuit of knowledge that we continue to push the boundaries of what is known, transforming the impossible into the possible, and shaping a future that exceeds our wildest imaginations. May our voyage into the unknown be a testament to the indomitable spirit of human ingenuity and our unwavering commitment to exploring the frontiers of the unexplored.

CHAPTER TWELVE

The Power of Imagination in Shaping the Future

I n the boundless realm of human potential, one force stands out as a catalyst for progress - imagination. From the earliest days of mankind, the power of imagination has driven us to explore new frontiers, challenge existing boundaries, and shape the world around us. In this chapter, we delve into the profound influence of imagination in shaping the future, particularly in the context of science, technology, and society.

Unleashing the Creative Mind:

Imagination, often regarded as the realm of artists, dreamers, and writers, is a powerful cognitive tool that transcends disciplinary boundaries. It allows us to envision possibilities beyond what exists in the present, paving the way for innovation and invention. As Albert Einstein once said, "Imagination is more important than knowledge. For knowledge is limited, whereas imagination embraces the entire world."

The creative mind at work is a tapestry of thoughts, experiences, and insights woven together by the limitless power of imagination. This innovative process relies on the ability to combine existing ideas in novel and imaginative ways. It involves embracing ambiguity, questioning assumptions, and daring to envision alternative paths and outcomes.

Throughout history, individuals who dared to dream big have brought about transformative change. From Leonardo da Vinci's envisioning of helicopters and parachutes centuries before their actual invention, to the imaginative leaps that led to the creation of the telephone, the light bulb, and the Internet, these breakthroughs were all born within the fertile landscapes of the human mind.

Fuelling Scientific Discoveries:

Imagination serves as an indispensable tool in the scientific process. At the forefront of scientific research, it sparks curiosity, drives experimentation, and propels us towards ground-breaking discoveries. Theoretical physicists, for example, rely heavily on their imaginative faculties to develop conceptual frameworks and formulate hypotheses that stretch the limits of our understanding of the universe.

Just as Galileo Galilei dared to challenge conventional wisdom and imagine a world where celestial bodies obeyed predictable laws of motion, so too did Isaac Newton propose a vision of the universe governed by gravitational forces. These grand intellectual strides reshaped our understanding of the cosmos and laid the foundation for future scientific breakthroughs.

In contemporary times, the power of imagination continues to shape our scientific pursuits. As technology advances exponentially, new doors of exploration are opened. From the hypothetical realms of quantum mechanics and the enigmatic nature of black holes to the intricacies of genetic engineering and artificial intelligence, it is the human imagination that propels us forward, allowing us to contemplate and explore previously unseen possibilities.

Furthermore, imagination plays a significant role in interdisciplinary scientific endeavours. Collaborations that bring together experts from diverse fields leverage the power of imagination to solve complex problems that require imaginative thinking. These efforts from different fields lead to big changes, like how nanotechnology, biology, and computer science are coming together in the field of bioengineering. This could lead to new ways of treating diseases and healing cells.

Manifesting Sustainable Solutions:

Imagination also plays a crucial role in envisioning a more sustainable future. As we grapple with pressing environmental challenges, it is our capacity to imagine alternative ways of living and interacting with our planet that can inspire transformative action. By visualising a world powered by renewable energy, where sustainable practises are the norm, we can strive towards creating a future where both humanity and the environment thrive.

Innovative minds across various fields are harnessing the power of imagination to tackle these pressing issues head-on. Architects and urban planners, for instance, are envisioning cities that harmoniously integrate nature, human well-being, and technological advancements, fostering sustainable living spaces that prioritise resource efficiency, active transportation, and biophilic design. They imagine skyscrapers adorned with vertical gardens, rooftop solar panels, and efficient water management systems, transforming concrete jungles into vibrant and sustainable ecosystems.

Likewise, engineers and scientists are leveraging their imaginative prowess to develop innovative solutions for clean energy generation, waste management, and climate change mitigation. From renewable energy sources such as solar, wind, and hydroelectric power, to advancements in battery technology and sustainable agricultural practises, imagination fuels the creation of practical solutions to address the world's sustainability challenges.

Moreover, the power of imagination extends beyond scientific and technological realms. It informs our social and political systems, giving rise to collective visions that shape the future of humanity. Imagination underlies movements for social justice and equality, challenging the present by imagining a future where discrimination and oppression no longer exist. It inspires policymakers to envision inclusive societies, where education, healthcare, and opportunities are accessible to all.

The Limitless Potential Ahead:

As we look ahead to the future, it is clear that the power of imagination will remain an indispensable force driving progress. With the advent of artificial intelligence, virtual reality, and other emerging technologies, the boundaries of what we can imagine and achieve continue to expand. It is through the convergence of human imagination and technological advancements that we can shape a future that surpasses our wildest dreams.

Imagine a world where artificial intelligence algorithms assist in diagnosing and treating diseases, revolutionising healthcare. Picture a reality where virtual reality and augmented reality seamlessly blend with our everyday lives, enhancing education, entertainment, and communication. Envision a future where gene-editing technologies open the doors to personalised medicine, eradicating genetic diseases, and extending human lifespans.

The potential for transformation through imagination is boundless. As we foster a culture that encourages creativity, curiosity, and open-mindedness, we unlock new dimensions of possibility. It is in these fertile mental landscapes that paradigm shifts occur, boundaries are shattered, and new realities are shaped.

Conclusion:

The power of imagination in shaping the future cannot be understated. It is the spark that ignites innovation, the force that propels scientific discovery, the compass that guides sustainable solutions, and the bedrock of social progress. As we continue to push the boundaries of what is possible, let us cherish and nurture this invaluable gift of imagination, for it is through our collective dreams and aspirations that we can reshape the world and create a better tomorrow. With each imaginative leap, we step closer to harnessing the full potential of humanity and shaping a future filled with endless possibilities.

CHAPTER THIRTEEN

Ethical Dilemmas in Science and Technology

E thics serves as the moral compass guiding the scientific and technological world. As advancements continue to reshape our society, it becomes increasingly crucial to explore and carefully consider the ethical implications accompanying these innovations. In this chapter, we will delve deeper into the most pressing ethical dilemmas encountered in the realm of science and technology, examining their complexities and potential solutions.

1. Balancing Scientific Progress and Potential Harm

The pursuit of scientific progress often presents a significant conundrum: the potential to improve lives versus the potential for unintended harm. Genetic engineering, for example, showcases this ethical quandary. While it opens up a huge range of possibilities, like curing genetic diseases or making people smarter or stronger, it also raises ethical concerns about things like playing God, genetically altering organisms without their permission, changing the course of natural evolution, and the wrong use of these technologies.

To navigate this ethical dilemma, scientists and policymakers must engage in thorough ethical discussions, seeking consensus and ac-

knowledging diverse perspectives. The establishment of robust regulatory frameworks and
ethical guidelines can help ensure responsible use of genetic engineering, safeguarding against unethical practises or misuse. Additionally, ongoing public engagement and
education are vital to foster understanding
and promote transparency in decision-making
processes.

2. Responsible AI Development

The field of artificial intelligence (AI) raises profound ethical dilemmas. As AI systems become increasingly sophisticated and
autonomous, there is a growing debate regarding their decision-making capabilities and the
level of human oversight. Questions emerge
about who should be held accountable for
AI-driven actions, the potential biases embed-
ded in AI algorithms, and the impact on em-

ployment, privacy rights, and social dynamics.

Developing responsible AI requires inter-disciplinary collaboration. Ethicists, scientists, policymakers, and industry leaders must work together to establish clear ethical principles, transparency standards, and accountability mechanisms. Efforts can include auditing and certification processes to evaluate the fairness, explainability, and safety of AI systems. Furthermore, promoting diversity within AI development teams can help mitigate potential biases and foster a wider range of perspectives.

3. Privacy, Data Protection, and Autonomy

In the digital age, ethical concerns about privacy and data protection have become increasingly significant. Rapid advancements in data collection, analysis, and storage have raised questions about how personal information is used, who has access to it, and the potential implications for individual autonomy and dignity.

Safeguarding privacy and autonomy requires comprehensive data protection laws, responsible data governance practises, and clear consent mechanisms. Striking a delicate balance between protecting individual privacy and promoting data-driven innovation is essential. Ethical frameworks that prioritise privacy by design, data minimisation, and informed consent can help navigate this terrain, ensuring that individuals maintain control over their personal information while enabling the benefits of technological advancements.

4. The Ethical Implications of Nanotechnology and Biohacking

Emerging technologies like nanotechnology and biohacking present unique ethical challenges. The ability to manipulate matter at the atomic level and enhance human capabilities beyond natural limits triggers moral dilemmas. Contemplation must be given to how these technologies are used, who has access to them, and the potential long-term consequences for equality, safety, and overall societal well-being.

To address these ethical implications, interdisciplinary dialogues between scientists, philosophers, policymakers, and the public are crucial. Establishing clear guidelines, regulations, and ethical standards to manage the development and utilisation of these technologies is essential. Ethical considerations must be underscored, ensuring that these advancements are used for the common good and that unintended consequences are carefully weighed and mitigated.

5. Technology's Impact on Social Inequality

While technology has the potential to bridge gaps and improve access to opportunities, there is a risk of widening social disparities. Ethical considerations must be given to ensuring that the benefits of technology are distributed equitably, addressing the potential consequences for marginalised groups, and actively working towards reducing digital divides.

To counteract social inequality, ethical frameworks should prioritise inclusivity, fairness, and social justice. This involves considering the societal impact of technological advancements, addressing the potential biases and discriminatory outcomes, and actively seeking ways to close the digital divide. Collaborative efforts between governments, tech companies, nonprofits, and communities can help ensure that technology is leveraged in

ways that promote inclusiveness and equal opportunities for all.

6. Environmental and Social Responsibility in Technological Development

The extraction and use of natural resources for technological advancements raise significant environmental and social ethics concerns. Balancing the needs of technological progress with the preservation of our planet and the well-being of communities affected by resource extraction becomes a complex ethical challenge.

Technological development must embrace responsible practises that prioritise sustainability and social well-being. This entails implementing sustainable manufacturing processes, reducing ecological footprints, and adopting circular economy principles. Furthermore, ethical supply chains that prioritise fair trade, community engagement, and minimising negative social and environmental impacts are crucial. By integrating environmental and social responsibility into technological development, we can ensure a more sustainable and equitable future for all.

7. Scientific Integrity and Responsible Research Practices

Maintaining scientific integrity and responsible research practises is fundamental to upholding ethical standards in the scientific community. Ethical guidelines and oversight mechanisms are necessary to ensure that research is conducted ethically, with consideration for the well-being of humans, animals, and the environment.

To promote scientific integrity, transparency, and accountability, robust peer-review processes and ethical review boards are essential. Responsible conduct of research includes adherence to ethical principles, proper handling of data, and the fair treatment of research subjects. Prioritising open and collaborative science can help foster trust and ensure that scientific advancements are pursued ethically, benefiting society as a whole.

Conclusion

The ethical dilemmas in science and technology are complex, multifaceted, and far-reaching. As we continue pushing the boundaries of knowledge and innovation, it becomes imperative to recognise and address the ethical implications of our actions. Striving for ethical frameworks that promote the common good, protect individual rights, and ensure the long-term sustainability of our society becomes a moral imperative. By embracing responsible practises, interdisciplinary collaborations, and inclusive decision-making processes, we can shape a future that upholds human values, equity, and the well-being of our planet.

CHAPTER FOURTEEN

Education and the Digital Revolution

The field of education has undergone a massive transformation in the digital age. The advent of technology has revolutionised the way we teach and learn, breaking the limitations of traditional methods and opening up new possibilities for students and educators alike.

One of the most significant impacts of the

digital revolution on education is the accessibility of information. With just a few clicks or taps, students can access a vast array of knowledge and resources from all around the world. The internet has become a treasure trove of information, allowing students to explore various subjects and topics in depth. Whether it's searching for scholarly articles, watching educational videos, or participating in online forums, the accessibility of information has democratised learning and empowered students to take charge of their education.

Moreover, the digital era has not only made information more accessible but has also transformed the very nature of knowledge. In the past, knowledge was mostly fixed and static, contained within the pages of books or dictated by teachers. However, the digital age has ushered in a new era of dynamic and interactive knowledge. Information is constantly updated and evolving, with new discoveries and perspectives being shared in real-time. Students can now engage with the latest research, collaborate with experts in the field, and contribute their own insights to the knowledge ecosystem. This shift from consuming knowledge to actively participating in its creation has transformed the educational landscape, nurturing a culture of curiosity, exploration, and lifelong learning.

The digital revolution has also brought about a fundamental shift in the way we engage with educational content. Traditional methods, such as textbooks and lectures, are no longer the sole means of instruction. Interactive learning experiences using multimedia presentations, simulations, and educational games have made the learning process more engaging and interactive. Students can now actively participate in their education, solving problems, conducting experiments, and exploring real-life scenarios. These digital tools not only facilitate a deeper understanding of concepts but also develop critical thinking,

problem-solving, and decision-making skills –
all essential for success in the modern world.

Furthermore, the digital revolution has fundamentally changed the dynamics of classroom learning. The traditional classroom is
no longer confined to physical spaces. Online
learning platforms and virtual classrooms enable students and teachers to connect and collaborate regardless of geographical locations.
Asynchronous learning, where students can
access materials and participate in discussions
at their convenience, has provided flexibili-
ty for those with busy schedules or diverse
learning needs. Moreover, online learning environments foster inclusivity by accommodat-
ing students with disabilities, providing closed
captioning, and offering alternative formats
for content.

The digital revolution has also given rise to
new pedagogical approaches. Blended learning, combining traditional face-to-face in-

struction with online components, has gained prominence. This approach allows for a more personalised and differentiated learning experience, catering to individual student needs and learning styles. It leverages the benefits of both online and offline interactions, creating a balanced and effective educational experience. Additionally, flipped classrooms have emerged as an innovative teaching model, where students engage with instructional content before coming to class, allowing in-person sessions to be dedicated to active learning, discussions, and problem-solving activities.

Beyond accessibility and engagement, the digital revolution has also impacted how educators teach and guide students. Teachers now have access to a wide range of digital tools and resources to enhance their teaching methods. They can create customised lesson plans, track student progress, and provide personalised feedback. Technology allows teachers to cater to individual student needs and adapt their teaching styles accordingly, fostering a more inclusive and effective educational experience. Moreover, technology supports collaborative learning, enabling students to collaborate on projects, engage in peer-to-peer discussions, and learn from each other's perspectives and experiences.

However, it is crucial to acknowledge the challenges and ethical considerations that arise with the digital revolution in education. The digital divide, where not all students have equal access to technology and the internet, remains an issue that educational institutions and policymakers must address. Efforts must be made to bridge this divide and ensure that all students, regardless of their socioeconomic backgrounds, have access to the tools and resources needed to succeed in the digital age.

Additionally, as technology becomes increasingly integrated into education, concerns regarding data privacy, cybersecurity, and the impact of technology on social interactions must be carefully managed. Educational institutions and policymakers should establish robust data protection policies to safeguard students' personal information and maintain students' online safety. Striking a balance between utilising technology for educational purposes and maintaining healthy face-to-face interactions is crucial, as social and emotional development is a vital aspect of education.

Furthermore, the digital revolution has accelerated the need to develop digital literacy skills. In today's interconnected world, students must not only consume digital content but also critically evaluate its credibility, understand the ethical implications of technology, and effectively communicate in digital environments. Educational institutions must prioritise digital literacy education, equipping students with the necessary skills to navigate and thrive in the digital landscape.

In summary, the digital revolution has transformed the field of education in profound ways, enhancing accessibility, engagement, and personalised learning experiences. It has opened up new dimensions for teaching and learning, allowing students to access a wealth of knowledge, collaborate with peers worldwide, and develop critical skills for the future. As we continue to embrace technology in education, it is essential to address the digital divide, protect student data, promote digital literacy, and strike a balance between online and offline interactions, ensuring a well-rounded and inclusive educational system for all.

CHAPTER FIFTEEN

The Future of Artificial Intelligence in Society

Over the past few decades, artificial intelligence (AI) has transcended mere sci-fi speculation and has become an integral part of our everyday lives. From voice assistants like Siri and Alexa to self-driving cars, AI is revolutionising various sectors and transforming the way we live, work, and interact with the world around us. As we look ahead to the future, the potential of AI is limitless, but it also raises important questions about its impact on society.

One of the most significant areas where AI will continue to exert its influence is in the workplace. Automation powered by AI has already started to replace certain job roles and will likely continue to do so in the years to come. While this may raise concerns about unemployment, it is crucial to recognise the potential for AI to create new jobs and enhance existing ones. By automating repetitive tasks, AI frees up human workers to focus on more complex and creative endeavours, ultimately leading to increased productivity and innovation. Moreover, the collaboration between humans and AI can lead to new and unprecedented levels of efficiency and productivity.

However, as AI becomes more advanced and capable of performing complex tasks, ethical considerations become increasingly pertinent. One of the key challenges lies in ensuring that

AI systems are developed and integrated into society in an ethical and responsible manner. This involves addressing issues such as data privacy, algorithmic bias, and accountability. It is essential to establish robust regulations and frameworks to govern AI development and deployment so that it aligns with human values and respect for individual rights.

Algorithmic bias is one area that requires urgent attention. AI algorithms are trained on large datasets, which can inadvertently contain biases inherent to the data. This can lead to discriminatory outcomes in areas such as hiring, lending, and law enforcement. Addressing algorithmic bias requires diverse and representative datasets, transparent algorithms, and ongoing monitoring to mitigate the potential for unintended discriminatory effects. It is the responsibility of developers and policymakers to ensure that AI is fair and unbiased, promoting equality and inclusivity.

On the societal front, AI has the potential to address some of the challenges that humanity faces. From healthcare to climate change, AI can play a pivotal role in finding innovative solutions. In healthcare, AI-powered algorithms can help in early disease detection, individualised treatment plans, and drug discovery. For example, machine learning algorithms can analyse medical images and patterns to detect diseases like cancer at an early stage, increasing chances of successful treatment. Similarly, in environmental sustainability, AI can aid in optimising energy usage, predicting weather patterns, and developing more efficient transportation systems. AI can analyse vast amounts of data to identify trends and patterns, enabling more informed decision-making for environmental initiatives.

However, it is important to approach the future integration of AI with a balanced perspective. While AI can bring numerous benefits, it is important to consider its limitations and potential risks. The growing reliance on AI systems raises concerns about dependency and vulnerabilities. Ensuring the robustness and security of AI systems becomes critical to avoid potential pitfalls and unintended consequences that could undermine public trust and safety. Preventing malicious use of AI technology, ensuring data privacy and security, and developing safeguards against AI-driven misinformation are vital for safeguarding society against potential harm.

Moreover, with the increasing influence of AI, the concept of "explainable AI" becomes crucial. As AI systems become more complex, opaque, and reliant on deep learning algorithms, it becomes challenging for humans to understand the underlying mechanisms and decision-making processes of these systems. This lack of interpretability raises concerns about the fairness, accountability, and transparency of AI systems. Researchers and developers are exploring ways to make AI systems more explainable, enabling humans to understand and trust the decisions made by AI algorithms. This is particularly important in fields such as healthcare and law enforcement, where decisions can have significant consequences on individuals' lives.

Education also plays a crucial role in shaping the future of AI in society. As AI continues to evolve, individuals need to acquire the necessary skills to adapt and collaborate effectively with AI technologies. This requires a focus not only on technical skills, but also on fostering creativity, critical thinking, and ethical decision-making. By equipping individuals with the right tools and knowledge, we can harness the potential of AI to empower individuals and foster a more inclusive and equitable society. Educational institutions and governments have a responsibility to equip individuals with the necessary digital literacy and ethical understanding to navigate a world increasingly influenced by AI.

Furthermore, as AI becomes more prevalent, the need for interdisciplinary collaboration becomes crucial. The development and integration of AI systems require expertise from multiple domains such as computer science, ethics, psychology, and policy making. Collaborative efforts between different fields can contribute to designing AI systems that align with human values, address social concerns, and promote sustainable development. Interdisciplinary research initiatives, industry-government partnerships, and inclusive dialogue involving multiple stakeholders are necessary for shaping the future of AI in a responsible and impactful manner.

The future of artificial intelligence in society holds immense potential to transform our lives for the better. From improving efficiency and productivity to tackling complex societal challenges, AI promises to shape our world in ways we are only beginning to comprehend. However, it is crucial to approach the integration of AI with careful consideration and attention to ethical implications. By addressing algorithmic bias, ensuring robust regulations, fostering interdisciplinary collaboration, promoting explainable AI, and empowering individuals through education, we can unlock the full potential of AI while safeguarding the values and well-being of humanity. Through responsible and thoughtful deployment, AI can become a powerful tool for societal progress, improving lives and creating a future that is both technologically advanced and socially responsible.

CHAPTER SIXTEEN

Sources and References For Further Reading and Research

The Evolution of Science and Technology:

1. "The Complete Dictionary of Scientific Biography". This is a key reference source for the history of science and technology. It con-

tains biographies of mathematicians and natural scientists from all countries and from all historical periods. "History of Science & Technology Collections, Articles, & Journals." n.d. Www.gale.com. https://www.gale.com/history-of-science-and-technology.

2. "The Oxford Companion to the History of Modern Science". This book provides a comprehensive overview of the history of modern science. Rider, Robin. n .d. "Research Guides: Introductory Research Guide for History of Science and Technology: Reference Works for History of Science & Technology." Researchguides.library.wisc.edu. Accessed November 18, 2023. https://researchguides.library.wisc.edu/c.php?g=178027&p=1168700.

3. The Evolution of Science, Technology and Innovation Policies: A Review of the Ghanaian Experience". This study examines the historical evolution of science, technology, and innovation policy in Ghana. Amankwah-Amoah, Joseph. 2016. "The Evo-

lution of Science, Technology and Innovation Policies: A Review of the Ghanaian Experience." Technological Forecasting and Social Change 110 (September): 134–42. https://doi.org/10.1016/j.techfore.2015.11.022.

4. "Improving Measures of Science, Technology, and Innovation: Interim Report". This report provides a comprehensive overview of the measures of science, technology, and innovation. In: JSTOR. 2016. "History of Science & Technology," October 1, 2016. https://www.jstor.org/subject/historyofscience.

5. "Transformations: Studies in the History of Science and Technology". This series seeks to capture the traditional history of science and technology, and emphasises explanations of the hard science practises that it seeks to historicise. Read "Improving Measures of Science, Technology, and Innovation: Interim Report" at NAP.edu. n.d. Nap.nationalacademies.org. Accessed November 18, 2023. https://nap.nationalacademies.org/rea

d/13358/chapter/11.

6. "The Elusive Transformation: Science, Technology, and the Evolution of International Politics". This book treats the roles of science and technology across the entire range of relations among nations, including security and economic issues. Almgren, Richard, and Dmitry Skobelev. 2020. "Evolution of Technology and Technology Governance." Journal of Open Innovation: Technology, Market, and Complexity 6 (2): 22. https://doi.org/10.3390/joitmc6020022.

7. "The Cultural Evolution of Technology and Science". This chapter explores how the principles and methods of cultural evolution can inform our understanding of technology and science. Both technology and science are prime examples of cumulative cultural evolution, with each generation preserving and building upon the achievements of prior generations. "Transformations: Studies in the History of Science and Technology." n.d. MIT Press. Accessed November 18, 2023.

https://mitpress.mit.edu/series/transformatio
ns-studies-in-the-history-of-science-and-tech
nology/.

The Power of Artificial Intelligence:

1. "Artificial Intelligence: A Modern Approach" by Stuart Russell and Peter Norvig. Pearson Education. October 2016. This textbook provides a comprehensive introduction to the field of artificial intelligence, covering various techniques and algorithms.

2. Goodfellow, Ian, Yoshua Bengio, and Aaron Courville. 2016. Deep Learning. MIT Press. This book offers an in-depth exploration of deep learning, a subfield of artificial intelligence that focuses on neural networks and their applications.

3. Sutton, Richard S, and Andrew Barto. 2018. Reinforcement Learning : An Introduction. Cambridge, Ma ; Lodon: The Mit Press. This book provides an introduction to rein-

forcement learning, a type of machine learning that focuses on learning from interaction with an environment.

4. Bostrom, Nick. 2014. Superintelligence. OUP Oxford. This book discusses the potential development of superintelligent AI and the possible consequences, risks, and strategies for ensuring the safe and beneficial development of such technology.

5. Russell, Stuart J. 2019. Human Compatible : Artificial Intelligence and the Problem of Control. [New York, New York?]: Penguin. This book explores the challenges of creating AI systems that are compatible with human values and can be controlled by humans.

6. Müller, Vincent C. 2020. "Ethics of Artificial Intelligence and Robotics." Stanford Encyclopedia of Philosophy. Metaphysics Research Lab, Stanford University. April 30, 2020. https://plato.stanford.edu/entries/ethics-ai/.

7.Domingos, Pedro. 2015. The Master Algorithm. Basic Books. This book provides an overview of the quest for a unifying algo-rithm that can learn anything and transform the world of AI.

8. Tegmark, Max. 2018. Life 3.0. Random House Us. This book explores the future of AI and its potential impact on society, addressing topics such as AI safety, ethics, and the future of work.

Communication Breakthroughs: From the Internet to Virtual Reality

1. Gleick, James. 2011. The Information. Vintage. This book provides a comprehensive history of information and communication technologies, from the invention of writing to the development of the internet.

2. Green, Lelia. 2010. The Internet. Berg. This book offers an accessible introduction to the internet and its impact on society, culture,

and communication.

3. Chan, Melanie. 2015. Virtual Reality : Representations in Contemporary Media. New York: Bloomsbury. This book explores the development and cultural significance of virtual reality, examining its applications in various industries and its potential impact on society.

4. Hanson, Jarice. 2016. The Social Media Revolution : An Economic Encyclopedia of Friending, Following, Texting, and Connecting. Santa Barbara, California: Greenwood, an imprint of ABC-CLIO, LLC. This encyclopedia provides an overview of the social media landscape, discussing the history, development, and impact of various platforms on communication and society.

5. Messaris, Paul, and Lee Humphreys. 2006. Digital Media : Transformations in Human Communication. New York: Peter Lang. This collection of essays examines the ways in which digital media has transformed human communication, focusing on topics such as social net-

working, virtual reality, and online privacy.

6. Grimshaw, Mark. 2014. The Oxford Handbook of Virtuality. Oxford: Oxford University Press. This handbook provides a comprehensive overview of virtual reality, covering its history, technology, applications, and potential future developments.

7. Spiegel, James S. 2017. "The Ethics of Virtual Reality Technology: Social Hazards and Public Policy Recommendations." Science and Engineering Ethics 24 (5): 1537–50. https://doi.org/10.1007/s11948-017-9979-y. This article discusses the ethical considerations and potential risks associated with the development and use of virtual reality technology, offering policy recommendations to address these concerns.

Medicine and Health: Revolutionising Healthcare:

1. Ausman JI. We need a revolution in medicine. Surg Neurol Int. 2011;2:185. doi: 10.4103/2152-7806.91140. Epub 2011 Dec 26. PMID: 22276239; PMCID: PMC3262996. The article argues for a revolution in the practise of medicine, emphasising the need to re-establish the doctor-patient relationship and take healthcare out of the control of the government.

2. Ahuja AS. The impact of artificial intelligence in medicine on the future role of the physician. PeerJ. 2019 Oct 4;7:e7702. doi: 10.7717/peerj.7702. PMID: 31592346; PMCID: PMC6779111. The article explores the potential uses of AI in medicine and considers the possibility of AI replacing or supplementing the role of physicians.

3. Weiner, Stacy. 2023. "5 Medical Advances That Will Change Patient Care." AAMC. May 4, 2023. https://www.aamc.org/news/5-medical-advances-will-change-patient-care.

4. Ellerbeck, Stefan. 2023. "5 Innovations

That Are Revolutionising Global Health-care." World Economic Forum. February 22, 2023. https://www.weforum.org/agenda/20 23/02/health-future-innovation-technology/. The article discusses how technological advances are starting to revolutionise the health-care sector, with innovations ranging from AI to gene editing.

5. Mahara G, Tian C, Xu X, Wang W. Revolutionising health care: Exploring the latest advances in medical sciences. J Glob Health. 2023 Aug 4;13:03042. doi: 10.7189 /jogh.13.03042. PMID: 37539846; PMCID: PMC10401902.

Energy and Sustainability in the 21st Century:

1. Tester, Jefferson W, and Aet Al. 2005. Sustainable Energy : Choosing among Options. Cambridge, Mass. ; London: Mit Press. This

book provides a comprehensive overview of sustainable energy technologies and their potential to meet global energy needs.

2. Boyle, Godfrey, and Open University. 2004. Renewable Energy. Oxford ; New York: Oxford University Press In Association With The Open University. This book offers an introduction to renewable energy sources, including solar, wind, hydro, and geothermal power, and discusses their potential for meeting global energy demands.

3. Vaclav Smil. 2016. Energy Transitions. Bloomsbury Publishing USA. This book examines the historical and contemporary transitions in energy systems, exploring the factors that drive these changes and their implications for the future of energy production and consumption.

4. ———. 2018. Energy and Civilisation : A History. Cambridge, Ma ; London The Mit Press. This book provides a comprehensive history of energy production and consumption, exploring the role of energy in shaping

human civilisation and the challenges of transitioning to sustainable energy systems.

5. Scheer, Hermann. 2013. The Energy Imperative. Routledge. This book argues for the urgent need to transition to 100% renewable energy sources and provides a roadmap for achieving this goal.

6. Apostol, Dean, James Palmer, Martin Pasqualetti, Richard Smardon, and Robert Sullivan. 2016. The Renewable Energy Landscape. Taylor & Francis. This book discusses the challenges and opportunities of integrating renewable energy sources into the landscape while preserving scenic values and promoting sustainable development.

7. Goodstein, David, and Michael Intriligator. 2017. Climate Change and the Energy Problem. World Scientific Publishing Company. This book explores the relationship between climate change and energy production, discussing the economic and policy implications of transitioning to sustainable energy systems.

Space Exploration and the Colonisation of Other Planets:

1. Roach, Mary. 2011. Packing for Mars: The Curious Science of Life in the Void. W. W. Norton & Company. This book explores the strange science of space travel, and the psychology, technology, and politics that go into sending a crew into orbit.

2. Kaku, Michio. 2018. The Future of Humanity : Terraforming Mars, Interstellar Travel, Immortality, and Our Destiny beyond Earth. Knopf Doubleday Publishing Group. This book discusses the possibility of humans living in outer space, the potential of immortality, and the prospect of ter-

raforming Mars.

3. Impey, Chris. 2015. Beyond: Our Future in Space. W. W. Norton & Company. This book discusses the history and future of space travel, the possibility of life on other planets, and the potential for humans to colonise the stars.

4. O'Neill, Gerard K, David Gump, Space Studies Institute, and Space Frontier Foundation. 2000. The High Frontier. Burlington, Ont. : Apogee Books. This book discusses the potential of space colonisation and the possibilities of living in space, including the design of space habitats and the challenges of space travel.

5. Davenport, Christian. 2018. The Space Barons. PublicAffairs. This book provides an inside look at the efforts by the private sector to colonise

and commercialise space.

6. Spudis, Paul D. 1996. The Once and Future Moon. Washington: Smithsonian Institution Press. This book discusses the scientific, commercial, and political benefits of lunar exploration.

7. Zubrin, Robert. 2000. Entering Space : Creating a Spacefaring Civilisation. New York: Jeremy P. Tarcher/Putnam. This book discusses the technical and economic feasibility of a spacefaring civilisation and the ways in which space exploration could benefit humans in the future.

8. Zubrin, Robert. 2021. CASE for MARS : The Plan to Settle the Red Planet and Why We Must. S.L.: Free Press. This book provides a detailed plan for the human colonisation of Mars, discussing the potential chal-

lenges and solutions for sustaining life on the planet.

9. ———. 2016. Mars Direct. Polaris Books. This book presents a practical plan for manned missions to Mars, discussing the technical and logistical aspects of such missions.

10. Wingo, Dennis. 2004. Moonrush. Burlington, Ont. : Apogee Books. This book discusses the potential benefits of lunar exploration, including the use of the Moon's resources for the betterment of life on Earth.

The Future of Transportation: From Flying Cars to Hyperloop:

1. Salo, Rudy. 2022. "More than a Sky-High Dream: Making Flying Cars Part of Our Future Transportation Infrastructure." Forbes. February 28, 2 0 2 2 . https://www.forbes.com/sites/rudysa lo/2022/02/28/more-than-a-sky-high -dream-making-flying-cars-part-of-ou r-future-transportation-infrastructur e/?sh=65b452814f4d

2. "The Flying Car—Challenges and Strategies Toward Future Adoption." Frontiers.   Accessed November 19, 2023. https://www.frontiersin.org/articles/ 10.3389/fbuil.2020.00106/full

3. Thomson, Freya. 2023. "Going Green: Transport Innovation in the Face of Climate Change." Open Access Government. January 18,

2 0 2 3 . https://www.openaccessgovernment. org/green-go-transport-innovation-el ectric-vehicles-climate-change/15152 1/

4. Bernhard, Adrienne. 2023. "What's Standing in the Way of the Flying Car?" Www.bbc.com. July 17, 2023. https://www.bbc.com/future/article/ 20230714-whats-standing-in-the-way -of-the-flying-car

5. Edwards, Chris. 2022. "Flying Cars and Hyperloops?" Eandt.theiet.org. December 5, 2022. https://eandt.theiet.org/content/artic les/2022/12/flying-cars-and-hyperloo ps/

6. "Innovating Sustainability: The Future of Transport." 2021. Www.valuer.ai. May 25, 2021. https://www.valuer.ai/blog/innovati

ng-sustainability-future-of-transport

7. Bayer, Eben. 2022. "Forget Flying Cars: Meet the Plane That's Really a Boat." Forbes. October 31, 2022. https://www.forbes.com/sites/ebenbayer/2022/10/31/the-future-will-not-have-flying-cars-but-who-needs-them/?sh=3bc6a807216d

8. Chris Stokel-Walker. 2020. "7 Major Breakthroughs in the Evolution of Sustainable Transport | Journey to Zero." Journeytozerostories.neste.com. December 13, 2020. https://journeytozerostories.neste.com/sustainability/7-major-breakthroughs-evolution-sustainable-transport#10f9606c

Exploring the Uncharted:

1. Briney, Amanda. 2020. "Discover the Age of Exploration." ThoughtCo. January 23, 2020. https://www.thoughtco.com/age-of-exploration-1435006. This article offers a concise history of the Age of Exploration, focusing on the period's significance in terms of global trade, wealth, and knowledge.

2. Editorial Staff. "5 Surprising Ways the Age of Exploration Shaped Modern Civilisation." Mental Floss. Last modified April 18, 2017.

3. O'Neill, Gerard K. 2000. Op. Cit.

4. Wohlforth, Charles P, and Amanda R Hendrix. 2017. Beyond Earth : Our Path to a New Home in the Planets.

New York: Vintage Books, A Division Of Penguin Random House Llc. This book examines the potential for human colonisation of other planets, focusing on the challenges and opportunities associated with establishing a permanent presence in space.

5. Davenport, Christian. 2018. Op.Cit.

6. Zubrin, Robert. 2019. The Case for Space : How the Revolution in Spaceflight Opens up a Future of Limitless Possibility. Amherst, New York: Prometheus Books. This book discusses the potential benefits of space exploration and colonisation, as well as the challenges and opportunities associated with these endeavours.

7. "Did the Age of Exploration Bring More Harm Than Good?" BBC History Magazine, BBC History Revealed & BBC World Histories -

HistoryExtra. Last modified January 25, 2019. https://www.historyextra.com/period /modern/age-of-exploration-bring-m ore-harm-than-good-americas-australi a-columbus-captain-cook/.

8. "The Age of Exploration." Encyclopedia Virginia. Last modified July 21, 2023. https://encyclopediavirginia.org/entri es/exploration-the-age-of/. This article provides an overview of the Age of Exploration, discussing the motivations, key figures, and impacts of this period in history.

9. https://www.mentalfloss.com/article /94520/5-surprising-ways-age-explor ation-shaped-modern-civilisation. This article highlights five lesser-known ways the Age of Exploration influenced modern

civilisation, including the emergence of the American cowboy and the introduction of smoking.

The Power of Imagination in Shaping the Future:

1. Bulley, Adam, Jonathan Redshaw, and Thomas Suddendorf. "The Future-Directed Functions of the Imagination: From Prediction to Metaforesight." Chapter. In The Cambridge Handbook of the Imagination, edited by Anna Abraham, 425–44. Cambridge Handbooks in Psychology. Cambridge: Cambridge University Press, 2020. doi:10.1017/978110858 0298.026.

2. Chang, Sharon. "Unadulterated

Wonder: Imagining the Future of
Imagination : Adjacent Issue 6." ITP
/ IMA. Accessed November 19, 2023.
https://itp.nyu.edu/adjacent/issue-6/
unadulterated-wonder-imagining-the
-future-of-imagination/.

3. Chimal, Abril. 2022. "Futures and the
Power of Imagination for
Transformation * Journal of Futures
Studies." Journal of Futures Studies.
October 5, 2022.
https://jfsdigital.org/2022/10/05/fut
ures-and-the-power-of-imagination-f
or-transformation/.

4. Hoch, C. (2022). Planning Imagination and the Future. Journal of
Planning Education and Research,
0(0). https://doi.org/10.1177/07394
56X221084997

5. Lewandowska, Kinga . n.d. "The Role
of Imagination in Projecting Your

Future." Intelligent Change. https://www.intelligentchange.com/blogs/read/the-role-of-imagination-in-projecting-your-future.

6. Moore, Michele-Lee, and Manjana Milkoreit. 2020. "Imagination and Transformations to Sustainable and Just Futures." Elementa: Science of the Anthropocene 8 (1).

7. "Imagination (Stanford Encyclopedia of Philosophy)." Stanford Encyclopedia of Philosophy. Accessed November 19, 2023. https://plato.stanford.edu/entries/imagination/.

Unveiling the Unknown:

1. Fornito, A., & Bullmore, E. (2015). Connectomics: A new paradigm for

understanding brain disease. European Neuropsychopharmacology, 25, 733-748. https://doi.org/10.1016/j.euroneuro.2014.02.011.

2. Groves, Robert. 2023. Unveiling Mysteries of the Universe. Balboa Press.

3. G. Deco and M. Kringelbach. "Great Expectations: Using Whole-Brain Computational Connectomics for Understanding Neuropsychiatric Disorders." Neuron, 84 (2014): 892-905. https://doi.org/10.1016/j.neuron.2014.08.034.

4. Prodan, L., Udrescu, M., & Vladutiu, M. (2006). A dependability perspective on emerging technologies. , 187-198. https://doi.org/10.1145/1128022.1128049.

5. Stieglitz, T. (2007). Restoration of neurological functions by neuropros-

thetic technologies: future prospects
and trends towards micro-, nano-,
and biohybrid systems.. Acta neu-
rochirurgica. Supplement, 97 Pt 1,
435-42. https://doi.org/10.1007/978
-3-211-33079-1_57.

6. Weaver, Stewart A. "4. The age
of exploration." Exploration: A
Very Short Introduction, 2015,
40-61. doi:10.1093/actrade/9780199
946952.003.0004.

The Power of Imagination in Shaping the Future.

1. Brennan, Patti. 2021. "Innovation
through Imagination — Envisioning

the Future of Technology-Supported Care." NLM Musings from the Mezzanine. September 15, 2021. https://nlmdirector.nlm.nih.gov/2021/09/15/innovation-through-imagination-envisioning-the-future-of-technology-supported-care/.

2. Bulley, Adam, Jonathan Redshaw, and Thomas Suddendorf. "The Future-Directed Functions of the Imagination: From Prediction to Metaforesight." Chapter. In The Cambridge Handbook of the Imagination, edited by Anna Abraham, 425–44. Cambridge Handbooks in Psychology. Cambridge: Cambridge University Press, 2020. doi:10.1017/9781108580298.026.

3. Canceran, Delfo C. "Social Imaginary in Social Change." Philippine Sociological Review 57 (2009): 21–36. htt

p://www.jstor.org/stable/23898342.

4. Chimal, Abril. 2022.Op.Cit.

5. Hoch, C. (2022). Planning Imagination and the Future. Journal of Planning Education and Research, 0(0). https://doi.org/10.1177/07394 56X221084997

6. Mills, C. W. (1959/1976). The Sociological Imagination. New York: Oxford University Press.

7. Schubert, Torben, Renée Eloo, Jana Scharfen, and Nexhmedin Morina. 2020. "How Imagining Personal Future Scenarios Influences Affect: Systematic Review and Meta-Analysis." Clinical Psychology Review 75 (February): 101811. https://doi.org/10.1 016/j.cpr.2019.101811.

8. "Harvard STS Programme » Research

» Platforms » Sociotechnical Imaginaries » Antecedents » Imagination in Science and Technology." Programme on Science, Technology and Society at Harvard. Accessed November 19, 2023. https://sts.hks.harvard.edu/research/ platforms/imaginaries/i.ant/imaginat ion-in-science-and-technology/.

9. "Imagination – The Cornerstone of Innovation." NLM Musings from the Mezzanine. Last modified September 14, 2021. https://nlmdirector.nlm.nih.gov/202 1/08/18/imagination-the-cornerston e-of-innovation/.

Ethical Dilemmas in Science and Technology:

1. "Emerging Ethical Dilemmas in Science and Technology." 2012. ScienceDaily. December 17, 2012. https://www.sciencedaily.com/release s/2012/12/121217162440.htm. This article provides a comprehensive overview of the ethical issues related to technology, including discussions on information technology, nanotechnology, and biotechnology.

2. Buckley, M.R.F. 2020. "When Science Can Harm." H ms.harvard.edu. March 2, 2020. https://hms.harvard.edu/new s/when-science-can-harm. This is a peer-reviewed journal that provides a platform for the analysis of ethical concepts and the ethical impact of new technologies.

3. Dae George, Richard T. 2007. The

Ethics of Information Technology and Business. Chichester: John Wiley & Sons.This book provides an in-depth analysis of the ethical issues related to information technology in business, including privacy, intellectual property, and cybercrime.

4. Ormandy EH, Dale J, Griffin G. Genetic engineering of animals: ethical issues, including welfare concerns. Can Vet J. 2011 May;52(5):544-50. PMID: 22043080; PMCID: PMC3078015.

5. Morris, Jonathan. 2006. The Ethics of Biotechnology. Philadelphia: Chelsea House Publishers. This book explores the ethical dilemmas posed by biotechnology, including genetic engineering and cloning.

6. Taylor, Michelle . 2019. "Top 10 Ethical Dilemmas in Science for 2 0 2 0 . "

Www.laboratoryequipment.com.
December 17, 2019.
https://www.laboratoryequipment.c
om/558920-Top-10-Ethical-Dilemma
s-in-Science-for-2020/.

Education and the Digital Revolution:

1. Akgun S, Greenhow C. Artificial intelligence in education: Addressing ethical challenges in K-12 settings. AI Ethics. 2022;2(3):431-440. doi: 10.1 007/s43681-021-00096-7. Epub 2021 Sep 22. PMID: 34790956; PMCID: PMC8455229.

2. Haleem, Abid, Mohd Javaid, Mohd Asim Qadri, and Rajiv Suman. 2022. "Understanding the Role of Digital

Technologies in Education: A Review." Sustainable Operations and Computers 3 (3): 275–85. https://doi.org/10.1016/j.susoc.2022.05.004.

3. H, Vipin. 2023. "The Future of Education: 5 ELearning Trends to Keep an Eye on in 2024." ELearning Industry. October 13, 2023. https://elearningindustry.com/future-of-education-elearning-trends-to-keep-an-eye-on-in-2024.

4. Michelle, Emmy. 2023. "The Future of ELearning: Trends and Predictions for 2023 and Beyond." ELearning Industry. August 20, 2023. https://elearningindustry.com/future-of-elearning-trends-and-predictions-for-2023-and-beyond.

5. Mitchell, Janel. n.d. "Technology and Ethics." Uen.pressbooks.pub . https://uen.pressbooks.pub/tech10

10/chapter/technology-and-ethics/.

6. Parker, Kim, Am, and a Lenhart. 2011. "The Digital Revolution and Higher Education." Pew Research Centre: Internet, Science & Tech. August 28, 2011. https://www.pewresearch.org/intern et/2011/08/28/the-digital-revolution -and-higher-education/.

7. Portwood , Nigel . n.d. "EDUCATION: THE JOURNEY towards a DIGITAL REVOLUTION Drawing on Insights and Research from around the World." https://global.oup.com/news-items/ OUP_DigitalReportFinal.pdf.

8. Taylor, Paul, and Kim Parker. 2011. "The Digital Revolution and Higher Education College Presidents, Public Differ on Value of

Online Learning Social & Demographic Trends." https://files.aeric.ed.gov/fulltext/ED524306.pdf.

The Future of Artificial Intelligence in Society:

1. Anderson, Janna, and Lee Rainie. 2018. "Artificial Intelligence and the Future of Humans." Pew Research Centre: Internet, Science & Tech. Pew Research Centre. December 10, 2018. https://www.pewresearch.org/internet/2018/12/10/artificial-intelligence-and-the-future-of-humans/.

2. Bajwa J, Munir U, Nori A, Williams B. Artificial intelligence in healthcare: transforming the practise of medicine. Future Healthc J. 2021

Jul;8(2):e188-e194. doi: 10.7861/fhj
.2021-0095. PMID: 34286183; PM-
CID: PMC8285156.

3. Capitol Technology University. 2023.
"The Ethical Considerations of
Artificial   Intelligence."
Www.captechu.edu.   Capitol
Technology University. May 30, 2023.
https://www.captechu.edu/blog/ethi
cal-considerations-of-artificial-intellig
ence.

4. Ilzetzki, Ethan, and Suryaansh Jain.
2023. "The Impact of Artificial
Intelligence on Growth and
Employment." CEPR. June 20, 2023.
https://cepr.org/voxeu/columns/imp
act-artificial-intelligence-growth-and
-employment.

5. "Impact of Artificial Intelligence (AI) on the Economy
& Jobs." n.d. Bank of Ameri-

ca. https://business.bofa.com/en-us/
content/economic-impact-of-ai.html.

6. Talty, Stephan. 2018. "What Will Our
Society Look like When Artificial
Intelligence Is Everywhere?"
Smithsonian. Smithsonian.com.
March 21, 2018.
https://www.smithsonianmag.com/i
nnovation/artificial-intelligence-futur
e-scenarios-180968403/.

7. Thomas, Mike. 2023. "The Future of
Artificial Intelligence." Built In.
March 3, 2023.
https://builtin.com/artificial-intellige
nce/artificial-intelligence-future.

8. UNESCO. 2023. "Artificial
Intelligence: Examples of Ethical
Dilemmas | UNESCO."
Www.unesco.org. UNESCO. April
21, 2023.
https://www.unesco.org/en/artificial

-intelligence/recommendation-ethics/
cases.

9. "What Are the Ethical Concerns with
   AI?" 2023. Www.cmich.edu. August
   21,   2023.
   https://www.cmich.edu/podcast/epis
   ode/what-are-the-ethical-concerns-wi
   th-ai.

www.ingramcontent.com/pod-product-compliance
Lightning Source LLC
Chambersburg PA
CBHW071246050326
40690CB00011B/2282